The Changing Mile

JAMES RODNEY HASTINGS
The University of Arizona

RAYMOND M. TURNER
United States Geological Survey

THE CHANGING MILE

An Ecological Study of Vegetation Change

With Time in the Lower Mile of an

Arid and Semiarid Region

THE UNIVERSITY OF ARIZONA PRESS
Tucson, Arizona

The University of Arizona Press
www.uapress.arizona.edu

Printed in the United States of America
21 20 19 18 17 16   10 9 8 7 6 5

ISBN-13: 978-0-8165-0014-7 (cloth)
ISBN-13: 978-0-8165-3525-5 (Century Collection paper)

L.C. No. 65-25019

♾ This paper meets the requirements of ANSI/NISO Z39.48-1992
(Permanence of Paper).

This book is affectionately dedicated to Bernie, a good climatologist gone wrong; to Jeanne, a frequently abandoned wife; to Mac, who started it all; and to Dick, who kept it going.

PREFACE

The University of Arizona's Institute of Atmospheric Physics has for some years sponsored a broad research program dealing with the basic climatology of dry regions; this study, reinforced by a similar interest of the United States Geological Survey in the hydrology of such areas, is a direct product of that program. Using materials drawn from a variety of disciplines, *The Changing Mile* explores the respective parts played by man and climate in altering the face of the arid Southwest of the United States and the arid Northwest of Mexico.

At the heart of the study is the notion of a "desert region," which, as used here, is not at all the same thing as a desert. To readers who live in a more humid region than that dealt with, or to those to whom "desert" means, in the European sense, an expanse of sandy wasteland, the distinction may not be obvious. But even in regions as dry as southwestern Arizona and northwestern Sonora, true desert is discontinuous. It makes up only one part—an important part to be sure—of a larger area with a considerable variety of climate. And although the desert itself is arid, other parts of the desert region may be only semiarid, or even subhumid. The desert region, then, includes the moist highlands that stud the desert, and those lying immediately adjacent to it. It includes the many microenvironments—streams, marshes, springs, roadsides, and the beds of dry washes—that may be wholly surrounded by desert, but are not themselves arid. To talk of the pines and oaks of the desert region is legitimate; they grow there in abundance. But to speak of them as being plants of the desert is inaccurate, because they can tolerate aridity hardly better than the forests of New England.

Recent changes in the vegetation of one desert region—southern Arizona and northwestern Sonora—constitute the subject of this book, and in determining what changes have taken place two approaches have been emphasized, neither of them an innovation. Historical records have been employed by a host of climatologists and geographers to recreate the vegetation and climate of the past; repeat photography has been similarly used, most recently

perhaps by Shantz and Turner (1958) and Phillips (1963).

The old photographs used here come from a variety of sources. To the people who clicked a shutter at the time and place right for our purposes, we are indebted; if we cannot always acknowledge that debt, it is because the intervening decades have not treated their identities kindly. Wherever possible the old photographs have been dated precisely; however, in cases where "about" is used, the true date may differ by as many as five years from that given. A list of the photographic stations, their locations, and their elevations appears in Appendix C.

For those stations located in the United States, elevations have been taken from USGS topographic sheets, and are given to the nearest fifty feet. For Mexican stations, elevations were determined barometrically, and are probably correct within one hundred feet. In all cases, the elevations apply to the camera station itself and not to the area pictured. Bearings for the line of sight were made by hand with a Brunton compass, and are given to the nearest 22-½°, *e.g.*, as north, north-northwest, northwest, or west-northwest.

For the new pictures a Crown Graphic camera and 2-¼ x 3-¼ Panatomic-X sheet film were used. In most cases, the camera was fitted with a 105 mm. lens, covering a field of about 35°; however, in the process of matching the new prints with the old, the field has frequently been reduced and may in some cases be as little as 20°. For some pictures, 65 mm. and 250 mm. lenses were used, the pertinent data for which may be obtained from the appropriate photographic manual (Morgan and Lester 1954).

In order to present a less formidable text to non-botanists, common names, wherever they represent real usage and are not merely an attempt at artificial "simplification," have been used in preference to scientific names. Although this practice puts at a loss the botanist who is not familiar with usage in the Southwest, there is no solution to the dilemma that will satisfy everyone, and the plant specialist, who has to cope with the problem more frequently than anyone else, is probably best able to bear the inconvenience. The common names have been employed rigorously; one and only one scientific equivalent exists for each common name used. Appendixes A and B list the equivalents and provide a way of getting back and forth between the two sets of terms.

Although the practice of Kearney and Peebles (1960) has usually been followed in botanical names, many departures have been allowed in the matter of the spelling and hyphenation of common names. For desert plants, for trees and shrubs, for cacti, and for grasses more specialized works have sometimes been used (Wiggins 1964, Benson and Darrow 1954, Standley 1920–26, Benson 1950, Hitchcock 1950, Gould 1951).

Acknowledgments. We are indebted to a number of persons for their assistance, and two of them—Bernard C. Hennessy and Stephen A McClanahan—have every reason to feel that this is as much their book as it is ours. Dr. Hennessy helped conceive the project and in the course of two pleasant summers did much more than help get it off the ground. Mr. McClanahan has been of continuous assistance in the darkroom and field.

James E. McDonald is due credit for the idea and for the initial historical research. Albert Macias, for two years a graduate assistant on this and related projects, has made many contributions. A. Richard Kassander, Jr., James E. McDonald, William D. Sellers, Christine R. Green, Clayton H. Reitan, Leon E. Salanave, Barbara Linthicum, and Lloyd G. Hamill, all members of the staff at the Institute of Atmospheric Physics of the University of Arizona, have been repeatedly generous in making available their time, knowledge, and skills.

Much of the work has been made possible through the interest and encouragement of members of the U. S. Geological Survey, particularly Luna B. Leopold, Thomas Maddock, Jr., and Richard C. Culler. With their unrelaxed support, the final stages of the field work were far more rewarding and the protracted chore of completing the manuscript was much less of an obstacle than would otherwise have been the case.

Part of the material was used by the senior author as a doctoral dissertation, and he is indebted to Russell C. Ewing and A. Richard Kassander, Jr., for guidance in that part of the work. The material is used here with permission of the Graduate College of the University of Arizona.

Charles H. Lowe, Jr.; Walter S. Phillips; Ray Brandes, Edith Kitt, and Sadie Schmidt of the Arizona Pioneers' Historical Society; Mildred Gholson; Albert C. Stewart; Doris Seibold; and George Olin, Naturalist, National Park Service, have been helpful in locating and supplying old pictures.

Charles T. Mason, Jr., has helped with taxonomic problems. Over the course of a long and pleasant association, Dean E. Mann has suggested many aspects of the social framework for the study of arid lands. Stanley Alcorn has collaborated on special problems relating to the saguaro.

In its initial stages, the study was partially sustained by the Rockefeller Foundation. The bulk of the work has constituted part of a larger project supported, in the case of the senior author, under Contract NR 082–191 with the Office of Naval Research and, in the case of the junior author, by the U. S. Geological Survey. James Hughes of the former agency has rendered personal encouragement as well.

Russell C. Ewing, Mario Rodríguez, and James E. McDonald have read all or parts of the manuscript and made various suggestions. Finally, Georgia W. Savage has labored patiently over a series of messy manuscripts, making many suggestions on matters of style, and forging consistency where there was very little.

JAMES RODNEY HASTINGS
RAYMOND M. TURNER

Tucson

TABLE OF CONTENTS

LIST OF FIGURES AND PLATES

The Changing Mile

INTRODUCTION

The eighties of the last century saw a series of changes initiated in the natural landscape over a large part of the region that had earlier belonged to the Borderlands of New Spain, but which at one time or another during the preceding forty years had been acquired by the United States.

By 1890 the old equilibrium had been so badly disturbed that signs of a major natural upheaval began to appear. Not quite overnight, but certainly during the course of a single summer, many of the streams of the region underwent a striking change in their hydrologic regimes.

Where the San Pedro River of southeastern Arizona formerly wound its sluggish course northward through a marshy, largely unchanneled valley, in August, 1890, it began carving a steep-walled trench through which it thereafter emptied rapidly and torrentially into the Gila. Where it formerly ran more or less consistently throughout the year, after 1890 its flow became intermittent, leaving the new channel dry over much of its length for most of the time (Hastings 1959).

With a few modifications much the same thing happened at about the same time along all of the major watercourses of southern Arizona, the Santa Cruz River, San Simon Creek, Babocomari Creek, Sonoita Creek, and the Sonoyta River.

Whether the hydrologic shift was confined to the Sonoran desert region; whether it occurred throughout the semiarid Southwest; whether, if occurring elsewhere, it followed the same chronology—these questions remain to be answered.

In the specific region with which this study is concerned—the northern Sonoran Desert and the highlands within it and to the east—irrigation ditches were left high and dry, and as washes adjusted their levels to those of the main streams, the valley floors became dissected and redissected. Much of the better farm land washed away; much of the rest was rendered unusable. To the farmer and the rancher the economic loss was severe, but not without at least partial compensation. Malaria, a major plague among the early settlers, disappeared along with the marshes, the beavers, and the fish.

Writing in 1903 about an area (pictured in Plate 38) on the west side of the Santa Rita Mountains, David Griffiths, Agriculturist with the Office of Farm Management, thought he detected change of another sort:

A close examination of the broad, gentle, grassy slopes between the arroyos in this vicinity reveals a very scattering growth of mesquite (*Prosopis velutina*) which is in the form of twigs 2 to 3 feet high with an occasional larger shrub in some of the more favorable localities. . . . One cannot tell whether this growth indicates that this shrub is spreading or not. The present condition rather suggests this possibility (Griffiths 1904: 29).

Seven years of additional observation confirmed his suspicion and he predicted that "the time is coming when these foothill grassy areas, which now have only an occasional small shrub will be as shrubby as the deserts and lower foothills . . . if not more so" (Griffiths 1910: 22).

Turning his attention to another plant, burroweed, he found that it too had "thickened and increased perceptibly during the last five years. . . . It is quite probable that the grasses unmolested would hold their own against its encroachment, but with the grassy vegetation weakened by grazing it may increase to such an extent as to crowd out nearly all of the valuable plants" (*ibid.,* 18).

In the same year J. J. Thornber, a botanist at the University of Arizona, agreed that "the mesquite is one of our species that appears to be on the increase." Writing in retrospect of the preceding thirty years, Thornber also called attention to the deterioration of the grazing ranges:

When once perennial grasses are killed out [by overgrazing and trampling] they are indeed slow to reassert themselves. Such denuded areas are claimed by

[3]

the less valuable six-weeks grasses . . . or worse yet, they are seized upon by one or more of the species of obnoxious weeds . . . unpalatable . . . for grazing purposes. On an area where practically everything else is grazed, they alone are left untouched by stock to continue reproduction, and to thus spread farther over the adjacent poorly grazed ranges. Unfortunately, too much of our once valuable grazing domain has become thus converted into unproductive weed wastes which hold the ground against valuable grazing plants (Thornber 1910: 276).

Also in 1910, an *annus mirabilis* in the perception of changing conditions, Forrest Shreve of the Desert Laboratory of the Carnegie Foundation noticed that things were not right with the saguaro, or giant cactus.

Young plants less than 1 dm in height are so rare, or inconspicuous, that nine botanists who have had excellent opportunities to find them report that they have never done so. . . . [A study of its establishment rate] compels the conclusion that it is not maintaining itself [in two favorable situations studied]. . . . A fuller knowledge of its germination and the behavior of its seedlings, together with a more complete knowledge of the periodicity of certain climatic elements within its range will be sure to throw light on the fall in its rate of establishment (Shreve 1910: 240).

In subsequent years the fact of a mesquite invasion has gained complete credence—although the reasons for it are still very much in dispute (Parker and Martin 1952). The failure of the saguaro to repopulate in some areas—most notably at Saguaro National Monument—has been well established (Alcorn and May 1962). One can read in various sources that burroweed and snakeweed have proliferated on overgrazed ranges (Benson and Darrow 1954).

Changes like these in the natural vegetation, together with others that have gone unnoticed, constitute the subject of this study. That they have taken place at all may be surprising. And that they have taken place on a scale so large over a period of time as short as eighty years is certainly surprising. Taken as a whole, the changes constitute a shift in the regional vegetation of an order so striking that it might better be associated with the oscillations of Pleistocene time than with the "stable" present.

To understand why this has happened is a problem of consuming interest. And Janus-like it faces in two directions. On the one hand, it is a challenging scientific problem; on the other, it is an important human problem. On the one hand, it is intimately involved with the workings of hydrology, climatology, and ecology. On the other, it is closely tied to man's activities both by its origins and its implications.

As a problem in human geography much of its importance stems from the climatic context in which the changes are set—that of an arid and semiarid region. With each decade, these drier parts of the earth's surface—variable and delicately balanced—become increasingly important; principally so by virtue of their being vacant.

For the most part the humid, temperate areas have undergone a relatively intensive development; to a considerable extent man's ability to find a place for still more of his kind lies in the degree of success with which he is able to populate the sixteen million square miles of the earth's surface where there is room, but where rainfall is inadequate for conventional agriculture (McDonald 1959: 3).

The utilization of these regions poses special difficulties, the paramount one being survival. An advanced civilization with abundant production and efficient transportation can evade the problem by attaching its arid parts to a humid, mother region that supplies them with food, energy, and most of their other needs. The arid zone in this case becomes a mere suburb and supplies only living space.

Few nations, however, outside of our own, can afford the luxury of a national bedroom, and with the majority the solution lies in skillfully using the meager resources at hand to develop an arid-land economy that, if not self-sufficient, can at least pay a large part of its own way.

Water is at the heart of the matter; how to use the meager supply effectively; how, if possible, to make it stretch. To realize that the dilemma as seen at a particular point in time may not be as confining as it seems, one has only to look back at the myth of the Great American Desert, the high plains lying between the ninety-eighth meridian and the Rocky Mountains.

Dr. Edwin James, physician with the Long Expedition of 1819–20 up the Platte and down the Canadian River, described the region as follows:

In regard to this extensive section of country, I do not hesitate in giving the opinion, that it is almost wholly unfit for cultivation, and of course uninhabitable by a people depending upon agriculture for their subsistence. Although tracts of fertile land considerable extensive are occasionally to be met with, yet the scarcity of wood and water, almost uniformly prevalent, will prove an insuperable obstacle in the way of settling the country (Thwaites 1905: 147, Morris 1926).

Yet in 1960 the tier of states carved from the High Plains ranked as major agricultural producers. In 1871 Colonel Green, Commandant of Camp Apache, Arizona Territory, described that country as follows:

If you wish any further correspondence from me as to my views of Arizona, I can only tell you I have been over a great portion of it . . . and found it a rocky, mountainous desert, not fit even for the beasts of the field to live in (*Arizona Citizen*, April 22, 1871).

Yet in 1960 Arizona had, excluding beasts of the field, a population numbering well over a million.

Walter Prescott Webb (1931), one of the few historians to treat at any length the problem of man's adjustment to aridity, notes that the Great Plains defied the Spaniard, held up the western migration of Americans for many decades, and fell at last only when the proper tools came to hand. In some cases these tools were new inventions—the windmill, barbed wire, the six-shooter. In other cases they emerged from an improving technology: new methods of dry farming; advances in plant and animal genetics. Finally, some arrived only when man, at times his own worst enemy, was able to modify the more cumbersome of his legal and social institutions.

The New England village could not be transplanted onto the Sonoran Desert; nor could the riparian common law of England governing brooks and rivers be applied in apportioning the scanty water of the West. No pithier commentary on man's institutional inflexibility exists than the *Report on the Lands of the Arid Region of the United States* of 1878 in which John Wesley Powell, father of the United States Geological Survey, pointed out, among other things, the absurdity of applying the philosophy of the Homestead Act—one hundred and sixty acres and a mule—to semiarid grazing lands (Powell 1962: 32f.).

That man needed to adapt his institutions to meet the conditions of the dry regions was Powell's central thesis. And to some extent the westerner did so: in developing the prior appropriation doctrine for water; in providing for irrigation districts so that farmers collectively could build the dams and the ditches that a single individual could not; in evolving, among the Mormon pioneers in Utah, a cooperative society flexible enough to respond rapidly to challenge.

But that, by and large, we have failed to adapt, and that mere technological brilliance obscures this basic failure is Webb's latest contention, in a sort of pessimistic afterthought to *The Great Plains*. Looking this time at the Far West (Webb 1957), he says that instead of adjusting, Americans have created an oasis civilization. Holing up in his city, the westerner lives a life of humid abundance. His wells tap the water resources of a vast surrounding area—and irrigate clover lawns. His air conditioners, powered by energy from Texas gas or from a river five counties away, enable him to evade the discomfort of high temperatures.

Efficient transportation links him with the other oases and with the humid East, but not at all with the hinterland around him. He is, to push Webb's thought possibly beyond what was intended, enmeshed with the national economy of abundance and not at all with the regional economy of scarcity. He lives at the expense of an arid region and surrounded

by it, but not with it. His technology enables him to escape its rigors without making concessions.[1]

Should an oasis need more water, engineers extend an aqueduct to a river two hundred miles away. Or they throw up surface reservoirs to collect runoff that has fallen at another place. Or they put down deeper wells to tap an underground supply deposited at another time, possibly during another geologic era. These are not, of course, solutions. They are mere borrowings from Peter to put off for a little the inevitable reckoning with Paul.

This study records the obverse of Webb's coin: what has been happening outside the oasis. Of necessity it is a record of man's failure to come to grips with aridity. He has failed because in his own scheme of values burroweed is a poor substitute for grass. A mesquite thicket in the terms of human economics is not the equivalent of a grassland. Not only does it waste water through excessive transpiration, it is relatively nonproductive.

But precisely where the failure lies is another question, the answer to which is contingent upon knowing what caused the streams and the vegetation to change in the first place. And this, as a scientific problem, is very much a matter of dispute.

In general the answers that are commonly given fall into two categories: those that hold man responsible, directly or indirectly; those that see natural factors, primarily climatic, operating independently of man.

If the cultural explanation is correct, our fault lies with having played bull in a china shop; in having, through inadvertence, brought down in ruins a delicately balanced structure that for all its sturdy resistance to heat, dust, and drought, was more fragile than it seemed to be.

If, on the other hand, the climatic explanation is correct, then we have merely failed to be sufficiently informed about the limits within which our environment operates. The error can have tragic consequences, even to an advanced society; but that the mistake is not new, even to such a society, the dust bowl on the High Plains in the thirties may attest to.

Evaluating the relative merits of the two explanations is by no means easy. Causation in historical, natural phenomena is a knotty problem for many reasons, not the least of which is that it places the experimental scientist, normally the one to be concerned with it, at such a disadvantage. He cannot go back and rerun the sequence of events, manipulating one variable at a time while he holds the others constant. Nor can he devise an experiment involving all the variables and perform it in any laboratory yet built. Too many factors must be considered, and the interaction between them is too complex.

So, although many experiments have been devised

from time to time to test specific hypotheses connected with one or another of the alleged agents of change, the conclusions, perfectly valid in themselves, are of limited application to the larger problem. More of the explanations will be examined later, but for illustration, consider three of them:

1) A rodent species feeds on succulent young saguaro seedlings. In the course of settling the Southwest, man exterminated most of the predators that kept the rodent population in check. Subsequently the number of rodents has increased to the point where very few saguaros ever get past the seedling stage uneaten. Man's coming, then, has indirectly resulted in the failure of the giant cactus to maintain itself.

2) Before American settlement, recurrent fires used to sweep the grasslands of the Southwest. These resulted in little permanent damage to the grasses, which were able to emerge the following season as strongly as before, but the burning did periodically kill off any woody seedlings that had become established since the preceding fire. The effect of fire prevention has been the proliferation of shrubs, and consequently the deterioration of the grasslands.

3) Cattle browse on mesquite beans, some of which pass unharmed through the alimentary tract to be deposited, still viable, in a medium ideal for their germination. The wholesale introduction of cattle into the Southwest has resulted in the wider dispersal of mesquite seeds and thus in the plant's spread.

In light of the one factor examined, each of these three statements presents an explanation for the one phase of vegetation change that it considers. The last two, at least, have wide credence; one frequently hears them advanced, for example, by ranchers to explain why their range land has been invaded by mesquite. But mesquite invasion is only one amid a host of changes that have taken place.

Do cattle also eat the seeds of ocotillo and turpentine bush—plants that, as the following pages will show, have registered substantial invasions of their own? And can these seeds also pass unharmed through the alimentary tract of a cow? The rancher does not know, and neither does the scientist, but the answer is probably "no." The cow-flap explanation breaks down when applied to the broader problem. So does the fire-suppression hypothesis in accounting for the widespread death among oaks and saguaros. So does an increased rodent population when one attempts to explain why *Acacia vernicosa*, whose seeds are a favorite rodent food, has increased, not decreased.

There is, of course, no reason to suppose that one cause has to suffice for all of the changes. The natural world being what it is, there is, in fact, every reason to suppose that many factors are involved. But this is all the more reason for not accepting single-factor ecology, even in the case where it attempts to explain changes in the distribution of only one species. For every striking relationship that can be demonstrated experimentally there may be three, untested, that in the complex environment of the plant counteract *the* reason. Prudence demands that any factor known to be operating in the case of one species at least be examined for its possible effect on others.

What must be done before a satisfactory consensus evolves in regard to the changing plant life of the Southwest is to look at each of the cultural explanations that have been suggested—not just the three cited—and test its applicability to each known change—not just one. And the various climatic hypotheses must be similarly tested. In spite of the affection with which paleobotanists regard climate, almost nothing is known about the extent to which a given change in rainfall or temperature can dislocate the range of a species.

When all of this has been done the conclusion—if anyone ever arrives at it—will follow from the summing up for each plant of a long series of big pluses and little minuses, zeros, little pluses and big minuses—one quantity for each factor, and generalizing from these, all the while juggling the complex interaction between the independent variables themselves.

The problem, handled experimentally, burgeons out of all proportion to the simplicity of the initial question: what caused the changes? And yet any other approach yields only another opinion.

This study makes no attempt to settle the matter once and for all, and in the last analysis it may raise more questions than it answers. But it does attempt to present a comprehensive view of the changes that have occurred since 1880, to sketch their historical context, to review the principal explanations that have been advanced to account for them, and to evaluate the evidence to date.

Inevitably some of the hypotheses considered fare better than others. For whatever consolation it may be to the losers, the explanations suggested here may not stand the test of time either. The moral perhaps is that the changes are "a better subject for study than for debate" (Anderson 1956: 776).

I

THE DESERT HABITAT

THE CONTEXT OF THE REGIONAL SETTING

The problem of "desert." The regional setting for this study is that of a "desert." The term is easy to define qualitatively as a dry region. A definition in terms of quantity comes harder, but is desirable to have since "dry" means different things to different persons, and even to the same person at different times.

Dryness connotes, first, a lack of precipitation, and a desert may be roughly defined as an area that gets less than ten inches of precipitation per year. But temperature considerations are also involved. Because there is less evaporation, a cool climate with less than ten inches of precipitation may support a more mesic vegetation than a hot climate with the same amount and may, in fact, give the impression of being only slightly arid.

A more elaborate approach, then, might define "desert" in terms of the ratio of potential evapotranspiration to rainfall (Thornthwaite 1948), or might, as with the widely known Köppen system or one of its modifications, use yearly mean temperature, yearly rainfall, and seasonal distribution of rainfall as a combined index (Trewartha 1954: 382).

Köppen's scheme as modified by Trewartha yields good results in roughing out the broad regional climates of the world and in relating them to major features of the earth's circulation. And sometimes it yields good results on a smaller scale: the boundary it delineates for the Sonoran Desert agrees fairly well with the line drawn by Forrest Shreve on the basis of vegetation.[1]

By and large, however, broad climatic classifications are of limited use and are apt to be popular only among geographers and those climatologists interested in broader phases of the subject. The ecologist in particular is likely to chafe because the resolving power of such classifications is low in the part of the spectrum with which he is most concerned (Daubenmire 1956, Muller 1947: 56). Where Köppen sees a desert, the ecologist sees a score of habitats, each with a distinctive plant and animal community of its own. The microclimatologist, whose interest lies with the turbulent and variable interface between earth and air, is apt to feel much the same as the ecologist: the regional desert has many microclimates.

The steepness of a slope, the plane on which it lies with respect to the sun, the reflectivity, or albedo, of a surface: these can modify considerably the small-scale climates over an area that has one homogeneous "temperature" when measured a few feet above the ground.

This multitude of microclimates, in turn, results in a multitude of plant communities. But the ecologist needs still further refinements because within one microclimate such nonclimatic factors as soil and topography may further diversify plant life.

The end result—as any student of the desert knows—is a multitude of little worlds, fragmented and varied, all existing within the larger framework of a "desert." The saguaro—the distinctive giant cactus of Arizona and Sonora—may grow densely on south slopes, but is seldom abundant on north slopes. In canyons receiving cold-air drainage, the plant life may be that of a distinctly moister, cooler zone. From a hill overlooking drier parts of the desert, the eye can follow the course of a wash by the tree growth along it. From a car the shoulders of a highway look foreign. Receiving runoff from the pavement in addition to their own rainfall, they belong floristically to a more humid habitat, and sometimes support a lush, "riparian" growth that contrasts sharply with the creosote-bush plain alongside them.

The quantitative definition of "desert," then, is partly a question of the scale on which one wants to look at an arid area, partly a question of which factors, and how many of them, one chooses for defining aridity. It seems unlikely that any set of requirements picked by the macroclimatologist will satisfy his colleagues interested in microenvironments. It seems unlikely, on the other hand, that the microclimatologist or the ecologist will come up with satisfactory quantitative definitions of their own. Some of the variables of most concern to them—soil moisture, for example—can be recorded only with elaborate apparatus. Others—temperature of the soil surface and potential evapotranspiration—are near-metaphysical concepts that defy physical measurement. For now a general quantitative definition

[7]

of desert much more precise than Köppen's is not feasible. And failing that, a desert is a dry region.

The world deserts. McDonald (1959) has given an excellent short discussion of the world's deserts and of the meteorology that dictates their existence. In general two kinds can be distinguished: (1) those deprived of an upwind source of oceanic moisture; (2) those that lie in a latitude chronically lacking in lifting processes.

The isolation from a moisture source may be physical—as with the Gobi Desert, surrounded on all sides by the Eurasian land mass. Or it may be topographically induced, as with a "rain-shadow" desert lying to the leeward of a mountain range.

Low-latitude west coast deserts, the second kind, owe their existence to the arrangement of the earth's wind systems and pressure belts. These arid regions cluster around the horse latitudes, zones characterized by high pressure cells in which dry, stable air masses from aloft sink and diverge.

To the north and south of the horse latitudes lie humid belts characterized by converging and rising air masses. But the deserts are too far poleward to benefit from the dynamic processes of the doldrums, even at the peak of their summer migration; too near the equator to be visited often by fronts and cyclones from the subpolar lows on the other side. And lying at the western edge of the continental land masses, they receive no benefits from trade winds or east-coast hurricanes.

The North American Desert. The area in North America falling within Köppen's definition of desert amounts to about 440,000 sq. mi., and as mapped by Shreve (1942a) appears in Figure 1. It straddles the boundary between the United States and Mexico and includes most of the peninsula of Baja California; major parts of the states of Arizona, Chihuahua, Coahuila, Nevada, Sonora, and Utah; lesser parts of California, Colorado, Durango, Idaho, New Mexico, Nuevo León, Oregon, San Luis Potosí, Texas, Utah, Washington, Wyoming, and Zacatecas. As world deserts go, it is not large. The fifth in size among them, it occupies about one-eighth the area of the Sahara and one-half the area of the Arabian (McDonald 1959: 7).

On the basis of climate and plant life it can be subdivided into four regional deserts. The Great Basin, the northernmost of these, has cool or cold winters with frequent snowfall. Its vegetation, dominated by sagebrush, has evolved from the primitive, temperate Arcto-Tertiary forest (Axelrod 1950: 286).

The Mohave Desert, next southward, lies in California, the Nevada wedge and a small part of northwestern Arizona; its elevated margins are delineated more or less by the range of the Joshua tree. Most of

the scanty precipitation falls during the cool winter season. The vegetation is somewhat transitional, having affinities both north and south. In recognition of one of its austral components the Mohave is sometimes lumped together with the two southern arid regions under the designation "creosote-bush desert."

The two southernmost of the deserts, the Chihuahuan marginally and the Sonoran principally, form the setting for this study.

The Chihuahuan Desert. Spreading out from the Mexican state after which it is named, the Chihuahuan Desert extends north into New Mexico, east into Texas and Coahuila, south into San Luis Potosí, Nuevo León, Zacatecas, and Durango. It occupies the northern end of the Mexican plateau, an elevated tableland bordered on one side by the Sierra Madre of the West and on the other by either lowland plains or the eastern Sierra Madre.

Like the Great Basin the Chihuahuan Desert is high: nearly half of it is over 4000 ft., and its upper elevational limit is well above 6000 ft. (Shreve 1942a: 236). In contrast to the Mohave Desert, and somewhat less to the Great Basin, the major rains occur in summer. The perennial vegetation can be characterized as "shrubby"—intermediate in appearance between the low, monotonous, "bushy" Great Basin and the varied, but highly arborescent Sonoran. It contains relatively few trees and few tall cacti, but has an abundance of smaller cacti, spiny shrubs, and succulent-leaved century plants and yuccas.

Separated from the main part of the North American Desert, the Chihuahuan presently makes contact with no other arid region. To the north and east it grades into the semiarid steppes of the Great Plains; to the west the Continental Divide separates it from the Sonoran Desert. At no point is creosote bush—the common denominator of the three southern deserts—continuous across the grassland that lies along the Divide and serves as a transition between the arid regions on either side.

The Chihuahuan should be only of passing interest here, where the concern is with an area farther to the west. But one interesting anomaly dictates otherwise. A large island of Chihuahuan Desert flora occurs along the San Pedro Valley of Arizona, well within the range of this study, and well away from the main body of the parent desert (Shreve 1942a: 235, Benson and Darrow 1954: 16).

Unmistakably Chihuahuan plants like tarbush, all thorn, mortonia, and *Acacia vernicosa* grow abundantly on limestone soils at elevations normally reserved in that area to the grasslands. As some of the following photographs will show, almost pure stands of grass did occupy them eighty years ago.

The Sonoran Desert. The Sonoran Desert, far and

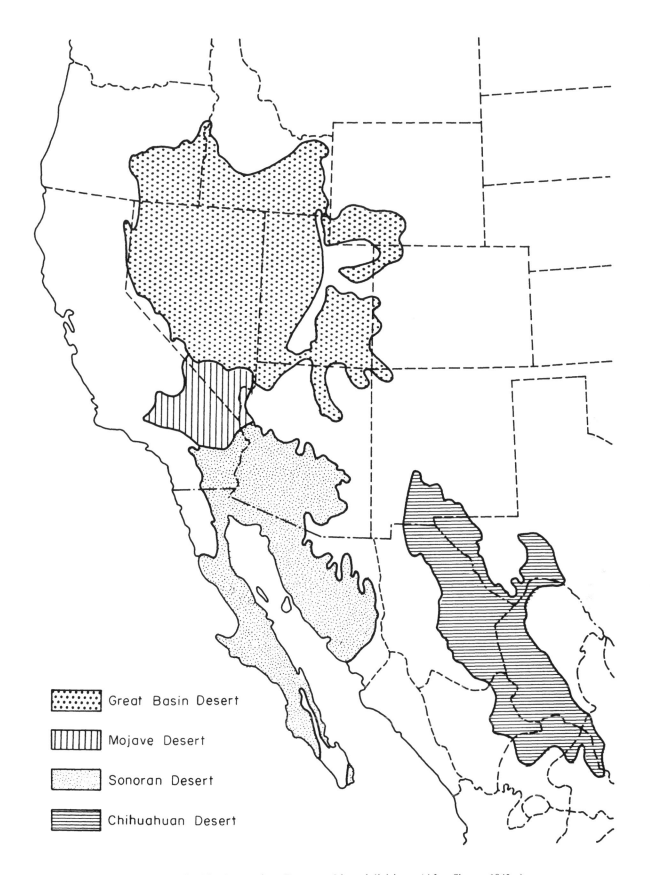

Great Basin Desert

Mojave Desert

Sonoran Desert

Chihuahuan Desert

away the most varied of the North American deserts, is also the lowest and hottest. It lies mainly in the Mexican states of Sonora and Northern Baja California, and in the territory of Southern Baja California, although a substantial section extends north of the International Boundary into Arizona and California.

The desert proper lies between sea level and about 3000 ft., the precise elevation varying with local soil and microclimatic conditions, and being higher to the north than to the south. Above this upper limit the isolated mountain masses that stud the region are in a very real sense wet islands anchored in a dry sea. For the Sonoran is not primarily the product of a rain-shadow. If it were, the plants of the mountain islands, except for modifications due to temperature, would resemble those of the desert itself. Atmospheric moisture is present over the region, occasionally in abundance,[2] and the orographic lifting induced when air flows up and over the mountain barriers compensates locally for the regional absence of dynamic lifting processes.

As a result of orographic lifting and rainout, the mountains are moist. Their tops may receive over thirty inches of precipitation per year and yet be located within ten miles of creosote-bush plains (Green 1959, Map 1).

The steep gradients of rainfall and temperature involved in this rapid transition from famine to feast find reflection in a sharp vertical zonation of the plant life. The interaction between climate and vegetation is so marked and so dramatic that the mountains make ideal places to observe the effect of climatic change on plant communities. Indeed, two of the vertical zones, which will be described later, are of as much interest here as the desert.

Besides the two sources already mentioned for the diversity of habitats to be found in the desert—localized microclimatic and edaphic differences, elevational zonation on mountains—a third must be recognized in connection with the extent of the Sonoran Desert, which stretches irregularly across twelve degrees of latitude, from above Needles, California, at thirty-five degrees south latitude to near San José del Cabo at the tip of Baja California at twenty-three degrees. Occupying an area of about 120,000 sq. mi., it is 870 mi. long and 400 mi. across at its widest point (Shreve 1964: 2).

Large-scale physiographic and climatic influences inevitably come into play over such an area. From west to east the land surface tilts upward, and the climate becomes increasingly continental with a larger daily range of temperature; a winter rainy season gives way to a summer rainy season; rainfall increases. North to south the hours of daylight, the temperature, the intensity and angle of sunlight all reflect the transition to lower latitudes.

Although a basic homogeneity of vegetation underlies the whole area—more perhaps from north to south than from west to east—several important characteristics do change, and largely in response to shifting climatic parameters.

THE CLIMATE OF THE REGION

The relation of desert vegetation to climate. Shreve (1964: 26) has defined a desert and has at same time described its implications for the vegetation:

It is essentially a region of low and unevenly distributed rainfall, low humidity, high air temperatures with great daily and seasonal ranges, very high surface soil temperatures, strong wind, soil with low organic content and high content of mineral salts, violent erosional work by water and wind, sporadic flow of streams, and poor development of normal dendritic drainage.

From the biological standpoint desert is best defined in terms of the limitations which have just been mentioned. For plants these serve to prevent the full degree of development that would enable them to form a closed covering, attain a considerable size, maintain vegetative activity throughout the year, and meet the environmental conditions without structural features or types of physiological behavior that tend to reduce their maximum performance.

In the case of the Sonoran Desert these general characteristics have to be qualified. Some plants do attain a considerable size—the saguaro, for example, or its even bigger relative, the *cardón*. Something resembling a closed cover can be found in mesquite bosques and the island mottes of the Foothills of Sonora (Shreve 1964: 31). At any time of year at least a few of the members of a community may be vegetatively active (Shreve 1964: 9). Finally, relative humidity is not uniformly low. Where the desert borders the Gulf of California, the higher moisture content of the air may be a factor in shaping such distinctive shore communities as the one pictured in Plate 94. Moreover, the possibility exists that some plants can make use of the heavy fogs that drift inland from the Pacific across the Vizcaíno region of Baja California.

Although qualifications like these temper the original statement, they by no means vitiate it. In one respect they emphasize the best feature of the definition by pointing out that this desert has special climatic and microclimatic characteristics that set it apart from others and that the plant life reflects these special conditions as faithfully as the generally xeric nature of desert vegetation reflects general aridity. Climate remains the single most important determinant for the plant life of an arid region, and to climate one must look to explain the uniqueness of the Sonoran vegetation: to precipitation, its amount, its variability, its spatial and temporal distribution; to temperature; to the various components of the heat balance.

The regional range of precipitation. Except for snowfall at higher elevations in the mountains, precipitation in the desert region falls mainly as rain. The average annual amount recorded varies from about 1.2 in. per year at Bataques (Hastings 1964a), west of the Colorado River in Baja California, to 19 in. per year at Ruby, in highlands surrounded by Arizona upland desert.

Few of the specific localities shown in the plates have reliable weather records, but a reasonable estimate of the range of precipitation that they encompass would be from about 18 in. per year at El Plomo Mine (Plates 1–3) to perhaps 5 in. per year in the Pinacate Mountains of Sonora (Plates 82–88), and perhaps 3.5 in. per year at Punto Cirio on the Gulf of California (Plates 92–94).[3]

But these annual amounts are averages and by no means indicate the water stresses that plants must undergo. In 1953 Yuma recorded 0.42 in. of rain, but in 1905, 11.41 in. (Green and Sellers 1964). San Luis, a Mexican station on the lower Colorado River, reported only 0.1 mm. in 1956, but received 152 mm. in 1927 (Hastings 1964b). Perennial plants must routinely be able to survive many weeks of high temperature without precipitation of any kind and with soil moisture values well below the wilting point. During June, rain fell at Yuma only seven times in the sixty-seven years from 1893 through 1959. In drier sections of the desert, droughts of seven to eleven months duration are not uncommon (Shreve 1944).

Coefficients of variation for precipitation. A useful statistical measure of temporal variability in precipitation is the coefficient of variation,[4] for which McDonald (1956: 6) has calculated some Arizona values and Wallén (1955: 67) some Sonoran. In general the coefficients vary inversely with the mean amount of precipitation received. For four of the stations lying in the area of greatest interest here the annual coefficients are 62 per cent at Yuma, Arizona; 42 per cent, Guaymas, Sonora; 30 per cent, Tucson, Arizona; 23 per cent, Atil, Sonora, northeast of Altar.

Yuma's mean annual rainfall over the period used by McDonald was only 3.25 in. Clearly a variation by 62 per cent or more from this scanty amount during one year out of three[5] represents a major fluctuation for the perennial vegetation to endure. Clearly too the population of annual plants must undergo large oscillations from year to year. During some years few may germinate, or mature; during other years the desert may approach lushness.

A factor which operates over much of the region to moderate the severity of the swings is the nonhomogeneity of the coefficients with respect to season. Defining summer as the six hot months (May through October), and winter as the six cold, McDonald has calculated seasonal coefficients. Contrary to what

might be expected, the spatially spotty, convective storms of summer, often less than a mile in diameter, are more dependable moisture sources from one year to the next than the large-scale, cyclonic disturbances of winter.

Whereas at Tucson the summer coefficient of variation is 40 per cent, for winter it rises to 54 per cent. Wallén's data for Mexico are not completely comparable since he has approached the problem by calculating another measure, the relative interannual variability,[6] for four key months, January, April, July, and October. It is nevertheless clear that for most of the Mexican part of the desert, the same pattern prevails. The relative interannual variability is much less at Atil and Guaymas for July than for other months (Wallén 1955: 72–76).

The implications of the seasonal variability for perennial plants are important. During the hot summer when conditions for vegetative activity are optimum for many species, but when soil moisture levels are low as a result of the arid spring, and water stress is greatest, the precipitation is most dependable from year to year.

This mechanism operates over much of the desert, but not all of it. Summer rainfall along the western edge, and particularly over the northwestern corner, is less dependable. In Yuma the summer coefficient of variation rises to 94 per cent, substantially larger than the winter coefficient of 75 per cent. In the northern and western reaches of Baja California the same is true, although in the eastern and southern parts of the peninsula the more general picture holds, and summer produces the minima (Hastings and Turner 1965). In the low, hot sink west of the Colorado River in Mexico, the most arid region in North America and, in summer, one of the hottest, summer variability is greater than winter. In this case particularly, where the plant life most needs dependability to mitigate aridity, it does not get it and must labor under an additional handicap. "Certain parts of [this region] have the thinnest plant covering to be found in North America" (Shreve 1934b: 373).

Seasonal distribution of precipitation. The apparent anomalies in summer coefficients over the western part of the region are directly related to factors governing the seasonal distribution of precipitation amounts. The rainfall of the Sonoran Desert is transitional in its regime between the Chihuahuan, with summer rainfall, and the Mohave, which receives the bulk of its precipitation during winter months. There are, in effect, two rainy seasons over most of the Sonoran Desert; the biseasonality must be considered a major factor in shaping the uniqueness of the plant life.

Shreve (1944: 108) illustrates its importance by citing three stations with nearly identical yearly amounts of precipitation, but different distributions.

Ensenada, on the west coast of Baja California, has a winter rainy season; Nogales, on the Arizona-Sonora border, has the Sonoran double-maximum, one coming in December–January, the other in July–August; Monclova, in eastern Coahuila, is characterized by summer rainfall. "The vegetation of the three localities is respectively chaparral, evergreen oak woodland and arid bushland. It is doubtful if a single native plant is common to the floras of Ensenada and Monclova."

The synoptic meteorology of the rainfall regimes. The synoptic features that produce three distinct types of desert within ten degrees of longitude have to do with the seasonal migrations of cyclonic systems and of two semipermanent high-pressure cells, the Bermuda high off the East Coast of North America and the Eastern Pacific high off the West Coast.

The winter rains are brought by migratory cyclones embedded in the westerlies as they make their annual migration southward. The Mohave and Sonoran Deserts profit marginally from these infrequent incursions of cold, rainy weather; the Chihuahuan Desert, on the lee side of the Sierra Madre, hardly at all.

With the northward advance of the sun in late winter and early spring the low-pressure systems retreat; their decreasingly frequent invasions herald the onset of a dry season which becomes more severe through spring and finally is either ended or reinforced by events related, this time, to incursions from the south.

These are the northward migrations of the two semipermanent highs. The Sonoran and Mohave Deserts, on the western continental slope, come under the influence of the Pacific high. The Chihuahuan Desert, on the eastern watershed of the continent, falls under the influence of the Bermuda high.

But here events lose their symmetry. In the Northern Hemisphere the western edge of an anticyclone is associated with converging, rising, unstable air; the eastern edge with sinking, diverging, stable masses from aloft. Over the deserts of the western slope the drought intensifies, giving rise to a dry, increasingly hot period appropriately called by Shreve the "arid fore-summer." Over the desert of the eastern watershed, however, the drought is dissipated by moist, southeasterly flow reaching progressively farther inland from the Gulf of Mexico as the high-pressure cell advances. May rainfall in Tamaulipas shows a sharp increase over April; June rainfall in Zacatecas, Durango, and Chihuahua, a sharp increase over May (Hastings and Turner 1965).

At the end of June an abrupt global readjustment brings a meteorological climax to Sonoran activities (Bryson and Lowry 1955) and provides one impor-

tant exception to the statement that western coastal deserts do not benefit from easterly activity. The highs move rapidly northwestward and enlarge. The Sonoran Desert is freed from the subsident eastern end of the Pacific anticyclone, which continues, however, to inhibit rainfall over the Mohave Desert and California. At the same time, the moist tongue at the western edge of the Bermuda high extends over the Continental Divide bringing air from the Gulf of Mexico. The Sonoran summer monsoon begins; rainfall over most of the Chihuahuan Desert continues to be abundant.

The implications of biseasonality. These are the principal synoptic events that underlie the seasonal distribution of precipitation and account for many of the differences among the three "creosote-bush deserts." The two-peaked distribution that prevails at the photographic sites has been characterized by Bryson (1957: 4) who appends to his description a provocative and pertinent notion:

Just north of Arizona, then, winter rainfall is more abundant than summer, but the double-maximum is dominant; just south of Arizona the double-maximum weakens and summer rainfall is dominant; just west of Arizona the summer rainfall disappears and winter rains dominate the southern California area; and just east of Arizona the winter rainfall loses *relative* importance and the summer peak dominates the annual march. Little wonder that through the centuries the ecology of the area, strongly controlled by moisture and its pattern of availability, has been apparently unstable.

This verbal description is reinforced by a map, reproduced in Figure 2, which shows his concept of Arizona as a southern biseasonal center from which gradients of monoseasonality radiate. The data are drawn from a Fourier analysis in which mean monthly precipitation amounts for Sonoran Desert stations have been fitted by least squares to two sine curves. The first curve has a period of twelve months and registers its highest amplitude in cases where there is monoseasonal precipitation with a single annual peak; the second has a period of six months and registers highest for stations with a biseasonal distribution containing equally spaced peaks. Figure 2 shows the *difference* in amplitude between the two harmonics. For positive values the amplitude of the annual wave is larger than that of the semiannual; for negative values, the reverse is true.

Except for Plates 89–97, the photographic sites are north of the zero line and lie in the area of biseasonal dominance. The southern boundary of the Sonoran Desert is approximately delineated by the +40 per cent isogram.

Much the same sort of picture arises from a simpler analysis involving the ratio of summer rainfall to winter, summer being defined as the six hot months, May through October, and winter as the six cold, November through April. In Figure 3 isograms

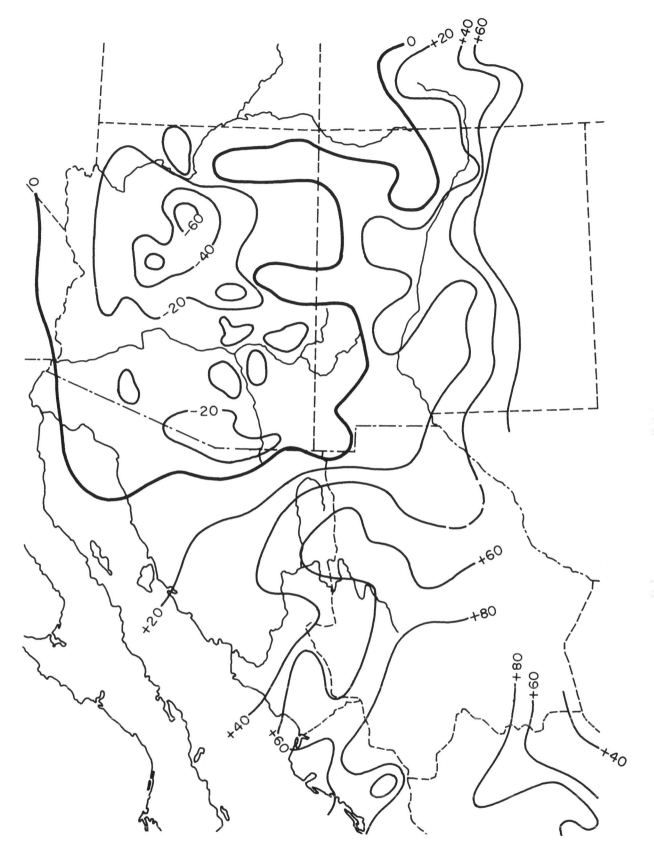

FIG. 2—Difference in amplitude in hundredths of an inch between sine approximations to annual and semiannual rainfall amounts. Positive: annual amplitude larger. (After Bryson 1957.)

[13]

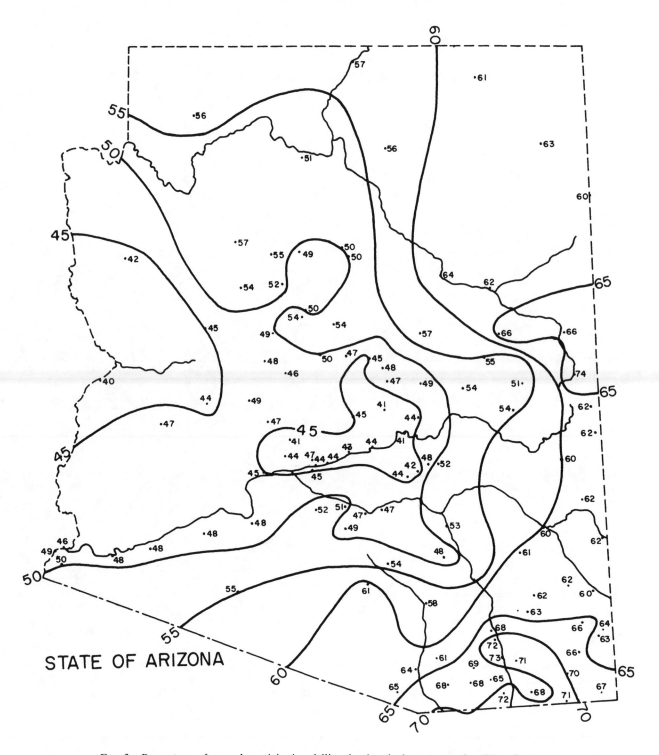

FIG. 3—Percentage of annual precipitation falling in the six hottest months (May-October).

for this datum have been plotted for Arizona; and their values, ranging from 45 per cent to 70 per cent in the desert region and the adjacent highlands, show not only the general biseasonal pattern but the lack of homogeneity within it, a characteristic that does not show up to advantage in Figure 2. Two prominent gradients appear: one extending from east to west, with winter rainfall becoming increasingly dominant; the other a similar trend from south to north.[7]

The first of these has been widely commented on. The north-south trend, which is even more pronounced, may be of equal importance, but has not been noted before. Many plants find their northern- or southernmost limits in this area of rapidly changing gradients; and the shifting patterns of seasonal precipitation may well play a part in terminating their ranges.

The north-south gradients are also of interest in connection with a problem closely related to vegetation change, the arroyo cutting that began along the streams of Arizona about 1890. The peculiar topography of the southeastern part of the state makes it one of the few regions in the United States that drain northward, and the valleys of both the San Pedro and the Santa Cruz Rivers sharply intercept the gradients of seasonal precipitation. In the course of one hundred miles the summer precipitation along them drops from over 70 per cent of the annual total to 50 per cent. At least one group of observers (Martin, Schoenwetter, and Arms 1961) has suggested that shifts in the seasonal distribution may be responsible for the onset of arroyo cutting, and while the tangible evidence is slight, the possibility nevertheless exists. The fact that channeling began on both streams near the 50 per cent isogram, where summer and winter precipitation are approximately in balance, and where a shift from one to the other might be expected to alter the corresponding balance between winter-dependent species and summer-dependent species, may not be significant, but deserves more investigation than it has received.

Effective precipitation. So far the discussion has centered around the patterns of actual precipitation. Of much more significance to the vegetation is effective precipitation, which can be defined as the fraction of actual precipitation that penetrates the soil surface far enough to become available to plants, without percolating on through.

Some water is lost by being intercepted before it reaches the ground. From the remainder the high temperatures of the desert exact their toll in evaporation. Still more is lost—or at any rate, redistributed—by virtue of the violence of summer storms, whose intensities may exceed considerably the rate at which the ground is capable of absorbing water, thereby producing runoff (Shreve 1934a).

Many of the resulting floods used to escape from the desert through the Gila, the Sonora, and the Yaqui Rivers. Storage dams and irrigation projects now capture most of this runoff; but the water is nonetheless lost to the native vegetation. Only that fraction can be considered effective that soaks into the gravelly beds along runnels, there creating moist, riparian microenvironments in which the less xeric native plants can flourish.

The efficient use of precipitation. The efficient use of that part of the precipitation that does survive as "effective" has also to be considered. The point has been made admirably by Benson (1962: 217–19) who cites the example of *Pinus ponderosa* var. *ponderosa,* a common western tree able to maintain itself in two greatly different environments.

The ponderosa pine forests of California grow in an area where the greater part of the annual precipitation falls in winter. Between this time and the time when it is warm enough for the trees to use the water, there is an appreciable interval during which a large part of the moisture is lost to the atmosphere through evaporation. Yet because winter precipitation is heavy in the region, soil moisture is still available when warm weather comes. The pines grow, and their seeds germinate before the onset of the long summer drought.

Forests of the same tree (although possibly a different ecologic race) occur in northern Arizona, but under a biseasonal regime of precipitation divided about equally between winter and summer. Because of the higher elevations of the Arizona pine forest, and because of the greater continentality of the inland climate, warm weather arrives relatively late in the spring, at a time when evaporation has already exhausted the moisture derived from winter storms. The trees undergo a period of enforced inactivity until summer, at which time the second rainy season begins. Temperatures then not only permit vegetative activity, but are optimum for it. "A high percentage of the water is available to the plants because there is no lag between [the period of rain] and a time warm enough for the water to be used" (Benson 1962: 217). Within the span of a few weeks vegetative growth occurs, the trees cast their seeds, and the seeds germinate.

In California an annual rainfall of twenty-five to thirty inches is necessary to support the forests; in Arizona an annual total of fifteen inches suffices.

The extent to which the same coincidence of rainfall and high temperature makes for the efficient use of summer precipitation on the desert is an interesting subject for speculation. The tendency among many ecologists has been to minimize the importance of summer rainfall to perennial plants on the grounds that evaporation consumes most of it. This is certainly not the case with some desert species, and may, in fact, not be true of many.

Recent experiments by the authors on the grounds of the Carnegie Institution's old Desert Laboratory [8] indicate that the efficiency of the saguaro in absorbing water varies directly with temperature, and that, given identical amounts of rainfall, the cactus is capable of picking up and storing a greater volume in summer than in winter. The shallow-rooted succulents of the Sonoran Desert may owe much of their adaptability to their ability to utilize summer moisture. Their abundance in a hot desert with biseasonal precipitation; their relative absence from a winter-rain desert (the Mohave), and from a cool, summer-rain desert (the Chihuahuan) may well rest upon this characteristic.

The influence of temperature. Plant responses to temperature are no less important than their responses to rainfall in determining which species can tolerate desert conditions; if anything, the ways in which a plant may react to temperature are more numerous and complex:

1) The temperature at which seeds germinate varies widely from one species to another, and with desert annuals is important in determining both their spatial distribution and whether they occur as winter or summer annuals, in response to winter or summer rainfall (Went 1949, Shreve 1964: 121).

2) In some cases a critical physiological function requires a specified temperature range. The case of the saguaro and its inability to pick up and store water in cold weather has already been mentioned. A similar limitation has been noted for the staghorn cholla, whose root growth is slow at temperatures below 20° C., and only optimum at 34° C. Its range is necessarily limited to places where the rainy season coincides with hot weather (Cannon 1916).

3) Within a single plant some processes may be most efficiently carried on at one temperature, other processes at another, so that appropriate diurnal and seasonal *variations* in temperature are necessary if the plant is to carry on all of its functions.

4) There is no evidence that unusually hot or cold weather can kill the mature, native plants of a region, however damaging it may be to exotics, and no matter how severe it may seem to human beings (Daubenmire 1957). Although a given occurrence may be rare in the weather records, the probability is high that in the many thousands of years over which the native vegetation has evolved, it has already experienced and withstood similar extremes. Nevertheless, the paleobotanical record makes it clear that *long-term* shifts in climate can displace plant distributions geographically, and can drastically alter the composition of the vegetation.[9]

The regional range of temperature. As might be expected, since they are determined primarily by large-scale factors like solar radiation and the earth's lag response, curves showing the annual distribution of average monthly temperatures on the Sonoran Desert look much the same for all stations. A well defined yearly maximum occurs at most places in July; a minimum in January. Inland the range between the two is largest, and the extremes of the curve tend to be most sharply defined—that is to say, the climate is most continental.[10] As a result of their proximity to water certain of the ocean and gulf stations—La Paz, Ensenada, Santa Rosalía, and Puerto Peñasco, for example—tend to have maxima that are delayed into August.

The driest part of the region—that lying around the head of the Gulf of California—is also the hottest. Average July temperatures at Yuma, Gila Bend, San Luis, Bataques, and Mexicali all exceed 90° F. Average January temperatures for the same group cluster around 55° F. (Green and Sellers 1964, Hastings 1964a and 1964b).

Of the photographic stations, the Pinacate Mountain region (Plates 82–88) and Guaymas (Plates 96 and 97) are probably the hottest. In the former the range of average monthly temperatures is from about 55° F. to 95° F.;[11] at the latter, from 65° F. to somewhat under 90° F. (Hastings 1964b: 53).

But as with precipitation values, means in temperature do not indicate the real stresses to which plants are subject, particularly when the standard exposures for thermographs and thermometers are made in weather shelters well above the ground. On a June day in Tucson, a relatively temperate desert station, Sinclair (1922) recorded a temperature of 161° F. at a depth of 4 mm. in the soil. At the same time a thermometer exposed in a standard shelter read only 108.5° F.

While the lesser figure may have more significance for a large plant, the greater is clearly more important to the germination of seeds and to the survival of seedlings. It—or the even harsher temperature at the soil surface—most nearly approaches the limiting condition for plant life.

In light of the stereotypes that prevail with regard to what desert conditions ought to be, the opposite extremes—those of cold—are even more surprising.

The distribution of freezing weather. There is probably no part of the Sonoran Desert that is free from periods when the temperature is below freezing (Turnage and Hinckley 1938: 547, Shreve 1934b: 379). Occasional winter cyclones penetrate as far into Mexico as the 20th parallel (Bryson 1957: 4, Page 1930: 1).

On one such occasion, the spectacular cold spell of January, 1937, observers reported frost damage as far south as Cedros, near the Sonora-Sinaloa boundary, where injury occurred not only to tender, subtropical plants, but to hardy types like canyon ragweed and soapberry, which range northward into the United States (Turnage and Hinckley 1938: 544).

Near Cedros the desert grades into subtropical thorn forest, and Shreve has observed that the boundary coincides with the beginning of the frost-free zone. He proposes implicitly, Turnage and Hinckley explicitly, that the line where freezing weather no longer occurs be considered the southern climatological limit for the Sonoran Desert.[12]

If one accepts this view, the mere occurrence of freezing can be dismissed as a factor making for diversification among desert plant communities, because it is a factor to which all of them are subject. But the frequency, intensity, and duration of freezing spells still remain of interest, and it seems likely that at least one of them may control the vertical zonation of plants on the mountain ranges within the desert.

The frequency and intensity of freezing weather. For the northern part of the region frequency is a relatively easy measure to examine, since Arizona data are available for the average number of days per year which have minimum temperatures below 32° F. Plotting these values against elevation for eighteen stations in the southeastern part of the state, where most of the zonation shown in the plates occurs, one gets the scatter shown in Figure 4a. There is, in fact, almost no relation to height above sea level. Bisbee at 5350 ft. has fewer than one-third as many freezing days as Willcox at 4200 ft.; fewer than half as many as Benson at 3635 ft. (Green and Sellers 1964).

Choosing "record low" as an easily obtained, if rough, approximation to intensity and plotting it in a similar manner, one finds a similar lack of dependence (Figure 4b). San Simon at 3608 ft. has experienced a low of −5° F.; Fort Grant at 4880 ft., +9° F.; Bisbee, 5350 ft., +6° F. It is clear that neither the number of freezing days, nor the intensity of cold bears much relation to elevation; therefore, to zonation.

A valuable study of freezing temperatures at three stations near Tucson sheds some light on this otherwise puzzling lack of correlation. For a period of five years Turnage and Hinckley (1938) maintained thermographs at two locations near the Desert Laboratory of the Carnegie Institution: the "hill" station, located on the shoulder of Tumamoc Hill, and the "garden" station at the foot of Tumamoc on the plain bordering the Santa Cruz River. Only one-half mile of distance and 330 ft. of elevation separated the instruments.

Yet the latter station on some nights recorded temperatures that were 20° F. lower than those at the former. For the five-year period beginning with the winter of 1932–33, the total number of freezing nights noted at the hill station was 38; at the garden, 263. The frost season, the period between the first and last freezes of a winter, averaged 36 days on the hill, and 157 days at the garden. During the winter of 1935–36 it extended only from January 2 to January 20 on the hill; from October 24 to April 6 at the garden, one-half mile away.

For two of the five winters a third station was maintained by an observer at Summerhaven, twenty-two miles away in the Santa Catalina Mountains at an elevation of 7600 ft. In general, minimum temperatures at Summerhaven and at the garden, 5100 ft. apart in elevation, were more closely related than minimum temperatures at the garden and hill, separated by only 330 vertical feet. During the winter of 1933–34 Summerhaven registered 57 freezing nights; the garden, 39; the hill, 6. The coldest temperature recorded on the hill during the same winter was 28° F.; at Summerhaven, 17°; at the garden 16°. "These data indicate the high mountains of southern Arizona do not experience minimum temperatures very much colder than . . . those of the desert lowlands" (Turnage and Hinckley 1938: 541).

The great differences between hill and valley, and the similarity between valley and mountain can be attributed largely to cold air drainage. The low humidity, clear skies, and long winter nights of the desert provide optimum conditions for rapid nighttime radiative cooling of the ground.[13] The air in contact with the soil surface cools by convection and conduction,[14] and the lower air temperatures propagate upward, producing an inversion condition in the lower atmosphere in which temperature increases with height, the reverse of daytime lapse conditions.

The air next to the earth, being coldest and therefore densest, tends to drain down slopes and away from high places into lower-lying areas, where it accumulates, contributing to a well developed inversion layer that may be several hundred feet deep in some desert valleys. In the Desert Laboratory experiment the hill station lay near the warmer top of such a "cold lake," [15] the garden at the bottom.

With this explanation in mind the scatter of Figures 4a and 4c becomes easier to interpret. The stations in the lower left of each diagram are located in valleys and the relatively great frequency of freezing temperatures in Figure 4a, and the relatively low mean temperature in Figure 4c can be attributed, as with the garden station, to cold air drainage.[16]

The cold "floods" resulting from such air flows have been recognized for many years as contributing to the interdigitation of life zones at the upper edge of the desert and above. Where two zones make contact the higher commonly reaches down into the lower along canyons and valleys, the paths for cold air drainage; the lower reaches into the upper along ridges. In some cases, trees may descend along streams as much as 3000 ft. below their lowest occurrence on north slopes.[17]

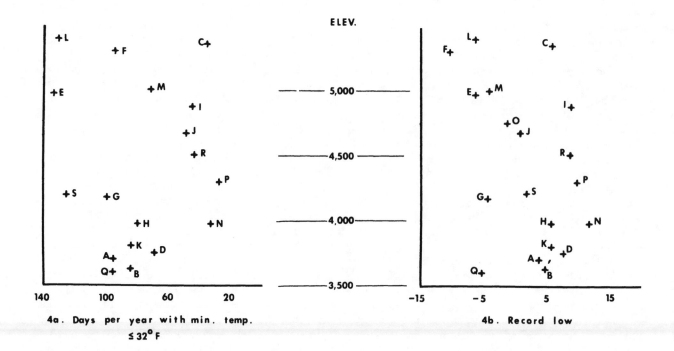

4a. Days per year with min. temp. ≤ 32°F

4b. Record low

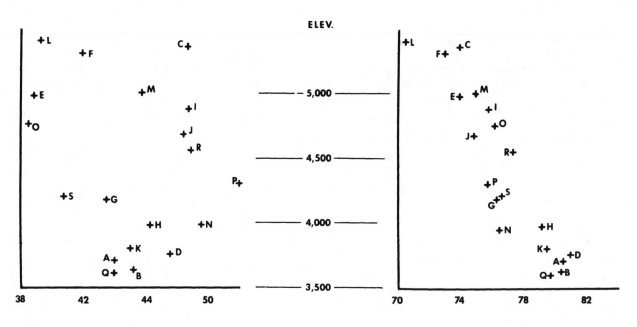

4c. Average daily min. temp.

4d. Average daily max. temp.

A-Apache Powd. D-Bowie   G-Cochise   J-Ft. Huachuca M-Portal   P Santa Rita
B-Benson   E-Canelo   H-Douglas   K-Nogales   N-Ruby   Q San Simon
C-Bisbee   F-Chiricahua I-Ft. Grant L-Ptd. Canyon   O-San Rafael   R Tombstone
                             S-Willcox

Fig. 4—The elevational distribution of various data of temperature for nineteen Arizona stations.

Considered in the context of an intermountain valley, rather than a mountain canyon, nighttime inversion layers may also account partially for the differences between the vegetation of the upper and lower parts of a bajada. On the part of the slope nearest the mountain one commonly finds in the Arizona Uplands such plants as saguaro, paloverde, and the frost-sensitive ironwood. Downslope this community grades into creosote bush and bursage which, in turn, give way on the cold bottom lands to mesquite. The transitions are usually attributed to soil (*e.g.* Yang and Lowe 1956), which typically ranges from coarse-textured and well drained near the mountain, to fine-textured and poorly drained in the bottoms. The soil factor may well be reinforced by inversion effects.[18]

The duration of freezing weather. In connection with zonation one further measure needs to be examined, the duration of cold, which need not be closely related to either intensity or frequency. Average 24-hr. minima, usually occurring at night (Figure 4c) do not show strong elevational relations, presumably because of cold air drainage. Average 24-hr. maxima, however, occurring in daytime after inversions lift, do (Figure 4d). Although Summerhaven and the garden station may experience approximately comparable nighttime freezing temperatures, the latter station with dawn rapidly proceeds to warm and to assume its appropriate position on the curve of temperature with respect to elevation. At the former station, freezing may persist well into the day, or even through it. The duration of freezing periods, then, may be substantially unrelated at stations with similar intensities or frequencies of cold.[19]

Since the patterns, because of cold air drainage, are quite irregular with respect to elevation, the intensity and the frequency of freezing temperatures evidently do not control the elevational limits of life zones. The duration of freezing spells, on the other hand, may be "the most potent cold temperature factor on the high mountain slopes of southern Arizona insofar as the absence of desert plants is concerned" (Turnage and Hinckley 1938: 538).

Some important facts about duration stem from elementary considerations about diurnal variation in winter. Daily maxima commonly occur about 1500 hrs., minima around 0600. Most freezing periods are limited to the hours of a single night, setting in after dark and terminating shortly after dawn, but in the event of a prolonged freeze, associated usually with an influx of Arctic air, temperatures below 32° F. may endure well into the next day. The freezing period can terminate with or before the diurnal maximum, in which case its duration will be limited to around 22 hrs.; or it can continue through the warmest part of the day, in which case it is intensified by the declining temperatures of evening

and night. Under these latter conditions, barring an unusual advection of warm air, the freeze will continue at least until the dawn of the following day and will result in a duration of 36 hrs. or more.

A frequency distribution for the length of freezing periods thus is discontinuous. There are many of 22 hrs. and less. There are some of 36 hrs. and more. Only rarely will there be intermediate values.

The evidence from experiments with small saguaros led Shreve (1911b) to conclude that between 20 and 36 hrs. of freezing temperatures were fatal to them. In their desert habitat, "the occurrence of a single day without midday thawing coupled with a cloudiness that would prevent the internal temperature of the cactus from going above that of the air, would spell the destruction of *Carnegiea*." Recognizing the existence of the quantum jump in durations, he suggested that the line along which it occurred should coincide with the northern limit of the saguaro.

Turnage and Hinckley (1938: 542–47) carried Shreve's generalization one step further: "The duration of freezing temperatures throughout a night, the following day and the following night . . . coincides with the northern limit of the Sonoran Desert and with the vertical limit of desert vegetation on mountain slopes and tablelands."

Although values for 36-hr. duration are hard to come by, another datum is readily available, the number of days per year when the maximum 24-hr. temperature does not rise above freezing. For reasons given above, the two measures are roughly equivalent.

In Figure 5 the latter datum has been plotted for all stations listed in *Arizona Climate* (Green and Sellers 1964). The line separating the locations which have never recorded such a day from those which have, coincides exactly with the northern boundary of the Sonoran Desert as drawn by Shreve (1964) on the basis of vegetation and flora.[20] It also agrees exactly with Shreve's distributional map for the saguaro (*ibid.*, 140).[21]

The climatological limits of the Sonoran Desert. Bearing in mind that "desert" is an imprecise term in both its climatological and ecological usage, and that in no event is there a sharp boundary between an arid region and its semiarid borderlands, fairly definite climatological limits for the Sonoran Desert may now be stated: on the south it is bounded by the line of farthest advance of frost; to the east and north, and on the mountain slopes lying within it, its limit is the line beyond which there is no midday interruption of freezing temperatures; to the west its extent is apparently fixed by an isohyet of summer rainfall.

Within these broad confines the climate is varied, yet homogeneous enough to produce a distinctive

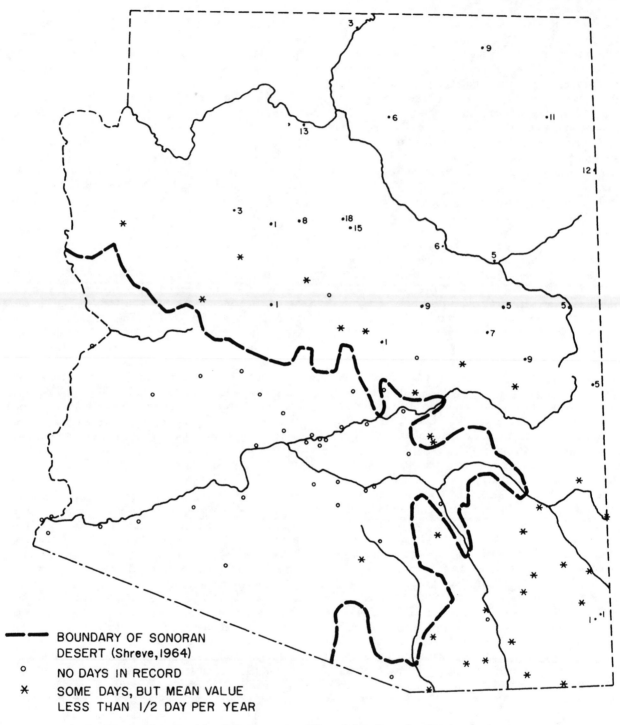

BOUNDARY OF SONORAN
DESERT (Shreve,1964)

○   NO DAYS IN RECORD

✳   SOME DAYS, BUT MEAN VALUE
     LESS THAN 1/2 DAY PER YEAR

FIG. 5—Mean number of days per year with no rise above freezing temperatures.

vegetation. The vegetation is homogeneous enough to be clearly recognizable; yet it also has a great deal of diversity.

THE VEGETATION OF THE REGION

Life zones. The plant life at any given place in the desert region depends upon its location with respect to axes running east-west, north-south, and vertically. Climatic gradients occur along all three: the vegetation tends to respond generally to the gradients, but with strong disruptions due to edaphic and microclimatic influences.

On the basis of the vegetation, and considering for a minute only the vertical axis, one can distinguish six "life zones" bounded by roughly horizontal surfaces that dip or rise with variation in rainfall and temperature, and in the case of at least one—the lower boundary of the oak woodland—that tilts toward the south. Whenever a land form intersects such a surface, the vegetation undergoes a pronounced, but usually gradual, change.

Several or, in some places, all of the zones may be seen in the course of ascending a single large mountain range. As one leaves the desert, the small-leaved plants, or microphylls, become less conspicuous, and plants armed with spines, thorns, or prickles are left behind. Grasses increase in importance, and, if areas of gentle terrain prevail at the upper edge of the desert, the grasses will predominate, with a local scattering of fleshy-leaved century plants or succulent-stemmed yuccas.

Beyond the grassland, widely spaced live oaks introduce the third zone. Many grasses from below are still present, but now occur as the matrix for tree members of the community. Higher still, the oaks, ever more closely spaced, are joined by other evergreen sclerophylls, or hard-leaved plants. The increased density together with the greater complexity of the flora characterize the upper edge of the woodland, and from it one passes to the fourth life zone, dominated by long-needled pines.

The lower portions of this zone still include several species of live oak, which extend upward beyond their own stratum of dominance and mingle among the taller trees in a wholly subordinant role. With increasing elevation, however, the vegetation of the forest becomes simpler; live oaks drop out, and pine species become fewer in number.

A fifth zone at yet higher elevation is signaled by the appearance of white fir and Douglas fir and the decreased importance of the pines. And finally several of the highest and largest mountain masses in the area support at their tops dense forests of spruce and fir.

Such is the vertical zonation that one finds in the regional vegetation: desert, desert grassland, oak woodland, pine forest, Douglas fir, and spruce-fir forests in order of increasing elevation. Only the first three zones, collectively comprising the first mile above sea level—the changing mile—will be dealt with here.

Of these three, the desert itself occupies the largest expanse. The grassland and the woodland are confined either to the upper reaches of isolated mountain ranges within the desert, or to the continuous highlands lying adjacent to it. Although the elevational limits of the three can be predicted reasonably well, the usual pattern may be altered somewhat by soil and topography. The grassland, for example, may be absent from certain areas; the desert then passes directly into woodland. Two zones may interdigitate at their line of contact, the upper reaching fingers down into the lower along canyons and other pathways for cold air; the lower into the upper along ridges and on south slopes. Thus where cool, moist, north slopes are in close proximity to dry, south slopes, there may be an inversion of the usual zonal positions (Shreve 1922: 270, Marshall 1957).

Latitude, the size of the mountain mass, its composition, and the elevation of the basal plain all play roles in determining the vertical limits within which the transition from one zone to another occurs (Shreve 1922: 271–73). Zonation, like other characteristics of the vegetational distribution, is variable and complex and subject to influence by a dozen climatic, edaphic, topographic, and cultural factors. Soil, temperature, and water are the most obvious controls, and by comparison the influence of living creatures may seem insignificant. Nevertheless, their activities over the centuries have helped shape both zonation and the other patterns.

During recent centuries, one such activity, at least in the premises of an anthropocentric world, has tended to overshadow the others. Man's role must now be examined.

II

THE INFLUENCE OF MAN:
INDIANS, SPANIARDS, MEXICANS

INTRODUCTION

The continuity of cultural factors. In attempting to assess the importance of man's role in changing the face of the desert region, the temptation is strong to postulate a "natural," static vegetation that existed one hundred years ago, and from this base proceed to estimate the extent to which Anglo-American settlement "disturbed" a previously "undisturbed" situation.

The simplicity of the model recommends it, but one suspects that, if for no other reason than its simplicity, it ought to be regarded skeptically. One also suspects that it partakes of some loaded value judgments about civilization and nature. In the past, the same emotional predisposition implicit in the model led Locke to postulate a "state of nature" as the origin of contractual, constitutional government before its usurpation by kings, Rousseau to blame society for corrupting the natural goodness of man, and Chateaubriand to compose his idyll of Réné and Atila in an unspoiled forest.

All of these attitudes have in common that they contrast a "natural" situation which is "good" with an "unnatural" influence that is "bad," and which, furthermore, stems from the activities of civilized man. They all involve certain assumptions about the nature and value of civilization that are not entirely tenable, and they all assume a process of historical development that does not accord entirely with the facts.[1]

Something of this tone carried over from the eighteenth century into the militant conservationism of the 1930's, and one still hears echoes of it from naturalists who feel that Anglo-American culture is "unnatural" and that it has been pervasively disruptive of an existing "natural" order.

One of the most vigorous critics of this view, James C. Malin, insists that the plow merely performs more completely and rapidly the same cultivation of the soil that takes place continuously in nature through the action of plants and animals; and he denies, for example, that water and wind erosion in the Midwest were the work of the white man. He emphasizes the "virtually continuous character of disturbance to soil and to vegetation during the whole [of] Pleistocene and recent time" (Malin 1956: 422–26).

Certainly long before the cattle drives out of Texas into the Sonoran Desert, man's activities played on integral part in the biological balance of this area. Mexican cattle roamed the grassland well before 1880; and Spanish cattle, long before the United States existed.

At least four centuries ago, and possibly eight, the Indian population of the desert was larger than the total population, European and Indian, that the area supported in 1880. It was probably as large, in fact, as the population of much of the region as late as 1940. Furthermore, the aboriginal peoples subsisted totally on the resources of the land without relying on food from a midwestern granary or from a warehouse in the central plateau of Mexico. Their impact on other living creatures, animal and plant, may have been substantial.

What we have to consider, then, is not a "natural" situation disrupted by "artificial" events following the American occupation, but rather a fluid environment shifting with the centuries under the impact of a succession of cultures, each differing somewhat from the preceding in its relation to the life around it. In the northern part of the region the Mexican phase gave way in the 1850's to the Anglo-American society that has since remained dominant. In some respects the new culture impinged more on its surroundings than did its predecessors, and in some respects, rather less.

The question to be answered is whether its different emphasis has been sufficient to account for any of the vegetative changes that have taken place since 1880.

THE INDIANS

Of the four general cultures to be considered—Indian, Spanish, Mexican, and Anglo-American, the first is least known. Partly this is so because it is the farthest removed in time; partly because it left no

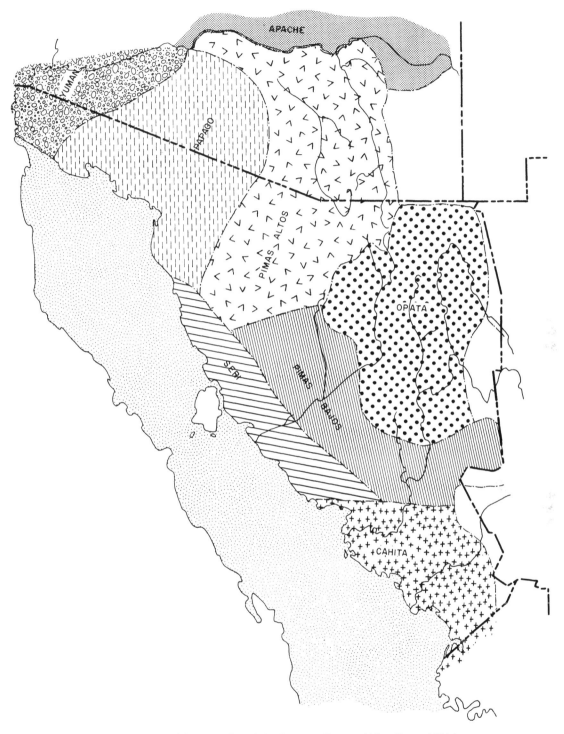

FIG. 6—Aboriginal peoples of the Sonoran Desert. (After Sauer 1934.)

written records of its own but must be recreated through archaeological remains, through second-hand impressions recorded by alien and sometimes unsympathetic conquerors, and through inferences made by anthropologists observing the Indian societies at a later stage.

By 1600 when Spanish contacts became frequent enough to result in a sketchy knowledge of the peoples inhabiting the Sonoran Desert, three groups may be distinguished, depending on the basis they enjoyed for their economic life (Spicer 1962: 8–15). The individual cultures making up these three groups are shown in Figure 6.[2]

Nonagricultural bands. Along the coast of the Gulf of California roamed nomadic bands of Seris, a loose Spanish designation for a number of bands who resisted Spanish attempts to settle them in mission centers, and who, practicing no formal agriculture, subsisted entirely by hunting, fishing, and gathering edible plant foods.

At the time of his work among the Seris late in the nineteenth century McGee (1898: pt. 1, 206 ff.) reported that cactus fruits from the saguaro, *cardón,* and prickly pear comprised about nine per cent of their diet. Flour made from mesquite beans, seeds collected from sedges and grasses in the mud flats along the Gulf, nuts gathered on the mountains, and small quantities of a local seaweed, in that order of importance, completed his list of their wild plant foods.

The band peoples. A second group that included the Apaches practiced a little agriculture and, although nomadic, had favored locations in which they recurrently settled. Although the Apaches are important to western history, by and large they are peripheral to an ecological discussion of the desert region. Like the Seris they hunted and were food-gatherers. Their formal agricultural activities, limited in extent, took place largely outside of the area considered here, but some of their informal practices—cattle stealing and raiding—later assumed considerable importance to it.

The ranchería peoples. A third group occupied by far the greatest part of the Sonoran Desert and may be called, after their diffuse, more-or-less permanent villages, the ranchería peoples. They subsisted by a mixture of farming and wild food gathering, with the former tending to predominate, but varying in importance among the different subgroups.

At one end of the scale the Yumans, occupying the valleys of the lower Colorado and the lower Gila, depended most heavily on wild foods and least on agriculture. To their south and east dwelt the Pimans, whom a blending of current and Spanish terminology customarily divides into three subgroups, the Papagos, the Upper Pimas, and the Lower Pimas. Their settlements, more townlike than those of the Yumans, still migrated, commonly moving in summertime to the valleys where there was surface water for irrigation.

The Papagos shared the occupancy of the arid lowlands with the Yumans and like that group tended to be simpler in their habits than the village peoples to the south. The Upper Pimas inhabited the uplands of south-central Arizona and northern Sonora; the Lower Pimas, the plains and the foothills of Sonora.

Along the southern and eastern edges of the desert lived the most highly organized of the ranchería groups, the Yaquis, and the Opatas. The former lived along the banks of the river whose name they share and whose perennial flow was large enough to maintain extensive agricultural activities and support permanent, well organized settlements. The latter inhabited the uplands around the Sonora, San Miguel, and Bavispe Rivers.

The size of the aboriginal popuation. Estimates of the size of the native population at the time of the Spanish contact vary within relatively narrow limits. Sauer (1935: 5) places the number of Lower Pima, Opata, Seri, Pima Alto, and Yaqui at about 155,000. Subtracting the part who lived outside the boundaries of the region being considered, and adding the Yuman peoples of the north, one can estimate that in 1600 the desert region, excluding Baja California, sustained a population of about 150,000. Using Spicer's figures (1962: 12, 99–100) for the same area at about the same time, one arrives at a smaller but comparable value of 120,000.[3]

In the wake of the social upheaval that followed the conquest, and under the impact of newly introduced diseases, the aboriginal population dropped precipitously.[4] Nearly three centuries of European colonization were required to restore the old density, and Sauer, writing in 1935 (p. 32) stated that for northwestern Mexico "aboriginal rural populations and present ones are much the same."

A demographic study of the Vizcaíno province in Baja California reached a more extreme conclusion:

After nearly 150 years in which the Central Desert has been open to settlement by people possessing the technological resources of western civilization, with substantially no native population to interfere with their activities, the population has slowly risen to only about one-fifth the number of the aborigines whom the missionaries found in the area a century earlier. There is no evidence that the number of residents will grow in the foreseeable future, and some indication that it will decrease. In terms of ecological adjustment, at least, modern Mexican civilization has not matched the achievements of the Indians who once occupied the Central Desert (Aschmann 1959: 268).

The ecological impact of food gathering. The extent to which the Indian population influenced the ecological balance of the region in pre-Spanish times cannot be stated with any precision. Certainly it impinged on the plant and animal life at many points.

An examination of the records for Baja California yields a partial list of the plants that were used as food by the Indians of the Vizcaíno: fruits or seeds from the organpipe cactus, cardón, sinita, prickly pear, barrel cacti, mesquite, ironwood, paloverdes, jojoba, canyon grape, elephant tree, and Mexican tea; the roots of yucca and Amoreuxia palmatifida; the stems, roots, and buds of century plants (Aschmann 1959: 78–93).

For northern Sonora Pfefferkorn (1949: 46–78), an eighteenth-century missionary at Atil, mentions, in addition to many of the plants already listed, the saguaro and mescal as foods; mangrove, ironwood, and mesquite as wood; creosote bush, yerba de pasmo, and a whole host of herbs gathered for their medicinal qualities.

From a twentieth-century report (Russell 1908: 68 ff.) on the Pimas of the Gila River Indian Reservation, several species of saltbush, catkins from the cottonwood, catclaw beans, the fruits of pencil cholla, staghorn cholla, and Mexican crucillo, desert cotton seeds, and Solanum elaeagnifolium can be added to the list of edibles.

Father Nentuig, stationed at Guasavas, Sonora, in the eighteenth century repeats most of Pfefferkorn's list, but includes amole and a group of herbs whose identities are not clear.[5] Missionaries in the oak woodland could doubtless have added plants from that zone. The list is lengthy but its ecological significance is not obvious.

Under some conditions food gathering may aid in the distribution of a species: the gathering of grass seed by flail and basket is one example (Carter 1950), the famous "second harvest" by the Seris and the Indians of Baja California, who recovered undigested seeds from fecal matter, may illustrate another.[6]

But when he wholly consumes and digests their fruits, man may interfere with the normal repopulation of plants. And where his attention is directed to parts other than the fruits, his influence may also be adverse.

McGee (1898: 207) cites the rarity of barrel cacti in Seri-land, and states that the few he saw were either wounded or dwarfed. Once one is past the sturdy spines and the waxy outer skin of the cactus, the pulp provides a ready source of water. In the dry reaches of the Central Gulf Coast, "its dearth suggests destruction nearly to the verge of extinction by improvident generations better armed with their hupfs and harpoons and shell cups than the subhuman beasts against whom the plant is so well protected" (McGee 1898: 207).

Aschmann (1959:79) speculates in a similar vein about the fate of century plants in the Vizcaíno province. He notes their absence from old mission sites and water holes and concludes that past disturbance by the Indians was responsible. "After finding . . . a clump an Indian woman could cut and carry enough hearts to feed her family for some time, but this procedure might entail the complete destruction of a plant community that was the product of decades, perhaps of centuries, of slow desert growth."

In the jungle of qualitative assumptions that must be made about the impact of food gathering, any quantitative information seems welcome. In his study of the Seri, McGee (1898: 214) reports that the 300 "full eaters" of that tribe consumed 27,000 lbs. of cactus fruits per year, or 216,000 individual fruits. Per person this amounts to 90 lbs. or 720 individual fruits.

There is, of course, no way of knowing what proportion belonged to each of the species involved, the cardón, the saguaro, and the prickly pears; in any event the proportion would vary from year to year with the harvest. If one assumes the saguaro's characteristics for all species, there are about 60 fruits per plant,[7] and an average plant density of perhaps 6 per acre.[8] On this basis a "full eater" would have consumed the output of about 2 ac.

If each of the 100,000 "full eaters" matched this performance, Indian food gathering would have accounted for the fruit from 200,000 ac., or about 400 sq. mi. When one considers the range of the saguaro, this is not an impressive total, and it becomes less impressive still when divided among several species of cacti, some of which have larger density values.[9]

In terms of the individual seeds consumed it is, of course, an enormous number: this many fruits are capable theoretically of producing 1.4×10^{10} new saguaros.[10] But a successful establishment is a rare event; virtually all of the saguaro's seeds perish regardless of man's activities. A more illuminating calculation is that one saguaro produces about 12×10^6 seeds during the century which constitutes its active reproductive life. In a stable stand one individual matures for each that dies. The end product of 12,000,000 seeds, therefore, is one plant that grows to maturity. Calculated on this basis, the total aboriginal consumption across the entire expanse of the desert resulted in the loss of only 1000 plants per year.[11]

Moreover, in view of the uncertainties of qualitative ecology one cannot be sure that food gathering tended, even this weakly, to interfere with repopulation. Many cactus seeds remain viable after passing through an intestinal tract (Alcorn 1961: 24), and the effect of human consumption might merely be their wider dissemination.

The ecological impact of cultivation. The artifi-cial cultivation of plants may have been a more potent factor in pre-Spanish ecology than food gathering. Many of the ranchería peoples practiced simple irrigation and there were probably few places amid the recesses of the desert that remained undisturbed for long if water and tillable land existed.

Where perennial rivers bisected the plains—along the Yaqui, the Colorado, and its tributary, the Gila—the opportunities existed for extensive agriculture. But most of the Sonoran streams were dry by the time they debouched from the uplands, and until the occurrence of the pump, irrigated farming was confined to their frequently narrow upper reaches.

Numerous clearings must have dotted the valley floor wherever the course of the stream left room between it and the rocky slopes at the side. As the river shifted with time or perhaps as the yield decreased from a plot, the fields may also have shifted, so that over the course of several centuries a continuous ribbon of disturbance can be envisioned, some parts actively cultivated, some recently abandoned, some overgrown.

An informative picture can be recreated of the conditions prevailing along the San Pedro River from the records of a trip in 1697 by the missionary Francisco Eusebio Kino (Bolton 1948, Mange 1926). His party found about 2000 Indians living in twelve villages scattered along the valley between Babocomari and Aravaipa Creeks. The approximate site of the fields of the southernmost of the rancherías, Santa Cruz de Gaybanipitea, appears in Plate 57. In 1697 the village contained about one hundred people, who practiced irrigation and tended a herd of cattle that Kino had given them earlier (Bolton 1948: I, 165). Below Santa Cruz about three miles lay Quiburi, the largest of the rancherías, populated by about five hundred Indians.

For some fifty miles thereafter the valley contained villages recently abandoned because of an internal war among the Sobaípuris. Below these, within a stretch of thirty-five miles, Kino found ten occupied rancherías. All of them practiced irrigation and raised corn, beans, cotton, and squash. Phrases like good pastures and fertile lands abound in the descriptions.

The history of Quiburi can be traced in some detail, and serves to illuminate both the mobility of the villages and the ease with which disturbance propagated along the watercourses. In 1692 the Quiburi Indians lived near present-day Redington at Baicatcan (Bolton 1948: I, 123; Bolton 1960: 364). On the occasion of Kino's visit in 1697 they were located at Quiburi. In the following year they moved to Los Reyes de Sonoidag on Sonoita Creek near modern Patagonia and downstream from the area shown in Plate 10 (Bolton 1948: I, 233; Bolton 1960: 385–86). By 1704 the old site at Quiburi had been reoccupied (Decorme 1941: II, 410). In 1762, faced by increasing Apache pressures, the Sobaípuris abandoned all settlements along the San Pedro (Nentuig 1951: 79), and the valley, except for brief periods, remained unoccupied until the coming of Anglo-Americans about a century later.[12] This history raises some interesting ecological points.

From the plates dealing with the San Pedro River (42–60) and Sonoita Creek (6, 7, 10) it is apparent that the native vegetation along the parts of their valleys that traverse the desert grassland has undergone major changes since 1880, an increase in the amount of mesquite being one. It is premature here to raise the involved question of why this species has proliferated, but it is worth noting that in spite of the disturbance along the two streams by aboriginal peoples, alternately cultivating and abandoning patches of land—disturbances which, ecologically speaking, resulted in a maximum opportunity for plant invaders to become established (Elton 1958: 147, Anderson 1956)—mesquite, though present along both valleys in sufficient quantity to provide seeds, did not become dense until recent times.

The Indians and fire. Omer N. Stewart and Carl O. Sauer have been in the vanguard of those asserting in recent years that the grasslands as broad, tree-and-brush free vegetational entities are the product of repeated fires set by aborigines in the course of their hunting activities.[13] The notion has gained wide attention and partial acceptance, and an early statement of the view by Sauer (1944: 554) forms an acceptable scientific hypothesis:

Too little is known about the great grasslands, and far too short a span of scientific observation, for us to be dogmatic about their origin. The . . . evidence . . . suggests that they are late features of plant geography, that they may have developed within the time of human occupation, that they may have been formed primarily by fire, and that the causes of the fires may have been man.

Humphrey (1958) and others (Branscomb 1956, Griffiths 1910, Thornber 1910) have applied the hypothesis to the desert grasslands of the Arizona-Sonora region; so inferentially, has Sauer by generalizing his earlier position to a view as extreme as Stewart's:

Wherever primitive man has had the opportunity to turn fire loose on a land, he seems to have done so, from time immemorial; it is only civilized societies that have undertaken to stop fires (Sauer 1956: 55).

The unrestricted burning of vegetation appears to be a universal trait among historic primitive peoples. . . . One may conclude that fire has been used by man to influence his geographic environment during his entire career as a human (Stewart 1956: 128–29).

It is not altogether obvious why the customs of all primitive peoples should be the same with regard to their use of fire; at least one anthropologist (Haury 1958: 72) familiar with the southwestern Indians states that "wholesale burning, argued by some to be a potent force in producing the grassland landscape, does not appear to have been a factor. . . ." Nor is it clear that the coming of civilized man need necessarily result in the suppression of fire. En route from Mexico City to Guadalajara in 1587, Father Ponce noted that "the savannas and pastures along the road had been fired, which is done so that new grass will come up for the sheep as soon as it rains, and for almost all the two and half leagues the Father Commissary was beset by smoke from both sides of the road" (Simpson 1952: 3). Since domestic grazing animals were introduced with the Spaniards, so, one concludes, was this particular occasion for burning. The incidence of fire on the Central Plateau, then, may have been greater after the coming of civilized man, not less. All in all, one doubts that the subject of aboriginally induced burning can support many of the generalizations that have been made, and perhaps it would be better to look at individual cases as they arise.

For the desert region there is evidence that two of the primitive peoples employed fire in connection with hunting. Pfefferkorn (1949: 198) notes that:

In various places in Sonora there are large areas covered with [sacaton]. This thick brush is infested with large numbers of rats and mice which the Sonorans sometimes hunt. Twenty or thirty and sometimes more Sonorans assemble and surround a given circle of brush. They start fires, setting the dry brush ablaze in a circle, and the animals hidden therein are forced to take flight. As the fire advances, the animals retreat more and more to the center and the Indians in turn close the circle on them.

He reports also that during June and July:

It is the custom at this season to burn the dried-out straw which remains lying on the field after the threshing. It often happens also that on their expeditions through the country the Apaches and Seris, as well as the herdsmen, light fires on the mountains to roast their meat. Because these are not extinguished before their departure, the fires spread easily and without resistance in the high grass, which is generally dried out by the heat of the sun at this time of year, seize on trees which stand in the way, and often cause a frightful conflagration. One thing and another fills the air with fiery vapors and increases the heat, which is great enough without this (Pfefferkorn 1949: 39–40).

These references establish that the Pimas burned the rubble in their fields—located in valleys—and used fire for flushing rodents out of swales of sacaton—a grass that occurs in valleys. Since Spanish accounts indicate that the desert valleys occupied by the Pimas of which Pfefferkorn writes already supported large amounts of mesquite (in contrast to conditions in the higher valleys of the desert grassland), one may question whether these burning activities were efficacious in suppressing shrubs, or, for that matter, even have any relevance to the question of maintaining the grasslands of the higher zone.

An account of the Apaches in 1796 by Don Antonio Cordero (Matson and Schroeder 1957: 343–44) describes the hunting of "deer, burro, antelope, javelina, porcupine, mountain lion, bear, wolves, coyotes, hare and rabbits."

At dawn a piece of terrain is encircled, which frequently is five or six leagues in circumference. The sign to commence the chase, and consequently to close the circle is given by smoke signals. There are men on horseback assigned to this project, which consists in setting fire to the grass and herbage of the whole circumference; and since for this purpose they are already placed ahead of time in their posts with torches ready which they make from dried bark or dried palmilla, it takes only a moment to see the whole circle flare up. At the same instant the shouts and the noise commence, the animals flee, they find no exit, and finally they fall into the hands of their astute adversaries. This kind of hunt takes place only when the grass and shrubs are dry. In flood season when the fields cannot be set afire they set up their enclosures by rivers and arroyos.

The reference establishes that the western Apaches used fire for general hunting purposes. The maintenance of brush-free grasslands, however, is not merely a qualitative question, but a quantitative one as well. If burning is an adequate explanation for the existence of grasslands, then it not only must have occurred, it must have occurred at frequent intervals over the entire expanse of the prairies. There is no evidence that it took place on the requisite scale with the requisite frequency. There is no evidence that it did not. One is justified only in stating that to the extent that fire did occur it may have helped locally to suppress woody plants. In short, it is by no means apparent that the desert grasslands of the Sonoran region owed their existence to "unrestricted burning" by aborigines, and one must agree with Sauer (1944: 554) that one ought not to be "dogmatic about their origin."

The ecological role of the Indian. The Indian cultures of the Sonoran Desert as they existed about 1600 were the product of many thousands of years of slow evolution.[14] They should not be thought of as emerging suddenly, nor once on the scene, being static; nor should their impact on the plant and animal life be regarded as fixed.

Vis-à-vis his environment the Indian occupied a fluid position. As the native societies evolved, growing larger and more complex and making greater demands on their habitats, his role changed. What the impact of the native peoples amounted to at its most intense, one can only guess. The effects of food gathering—if significant at all—were probably geographically localized around villages. Hunting may indirectly have been important to the vegetation at an early date if, as some authorities believe, it resulted in the extinction of Pleistocene grazing mammals. Farming must have been a potent source of disturbance, but was confined principally to the valleys where there was surface water. About the Indian's use of fire in hunting, the evidence is scanty, but the safest assumption is that he added little to the natural incidence of burning.

The Indian's role in 1600 was already an anachronism. After 1519 when Hernando Cortés dropped anchor off what is now Vera Cruz, a new series of events brought other forces stirring. The desert hardly felt them for seventy-five years. Then before the rising winds, the ecological mobile shifted slightly and its parts found a new equilibrium.

THE SPANIARDS

The age of exploration. By 1521 Cortés' maneuvers had accomplished the fall of the Aztec empire and the Spaniards, using Mexico City as a base, raced out in all directions across the new land. Within ten years Nuño de Guzmán founded Culiacán, 700 mi. away from the fresh ruins of Tenochtitlán. His slavers, raiding among the Yaqui, made the initial penetrations of the Sonoran Desert.

By 1536 Alvar Núñez Cabeza de Vaca had journeyed across Texas and down the Río Sonora where he was hospitably received by the Opatas.

In 1537 Fray Marcos de Niza retraced in part Cabeza de Vaca's route and saw, he said, the cities of Cíbola. Two years later Francisco Vázquez de Coronado—trailing sheep, horses, and cattle in his wake—followed the Friar, descended the San Pedro, crossed the highlands of east central Arizona, and found to his disgust not the fabulous cities, but the mud pueblos of New Mexico. One of his lieutenants, Melchior Díaz, crossed the Sonoran Desert from the Río Sonora to the Colorado River and partially explored the upper reaches of Baja California before falling victim to his own lance. Alarcón, cooperating with Coronado by sea, reached the Colorado and proceeded up it possibly past the mouth of the Gila.[15]

Spectacular though they were, the conquistadors impinged very little on the life of the desert, and the same can be said of the explorers who followed them at intervals during the remainder of the sixteenth century. Not, in fact, until the end of the century when missionaries quietly reached the Yaqui River did the Spaniard make his presence felt.

The mission. Jesuit activity on the northwestern slope dates from about 1591 with the work in Sinaloa of Gonzalo de Tapia. The Sonoran Desert peoples fell under its influence in 1617 when Andrés Pérez de Ribas (1645) led an entrada to the Yaqui River, and thereafter missionary activity spread rapidly across the region.[16]

The Jesuit influence was profound out of all proportion to the small number of Europeans involved. It wrought nothing less than a revolution in the cultural and economic life of the Indians and considerably modified their ecological role. The unique institution through which it operated was the mission, a self-sufficient unit that besides accomplishing the religious aims of the Church, played a key part in the imperial policy of Spain (Bolton 1917).

Besides spreading Christianity, the mission acted as an agent of peaceful conquest; in addition, it served as a tool of acculturation and assimilation. In theory it existed only on the frontier, and in theory its lifetime was brief—ten years. During that period it was expected to concentrate the Indians in settlements near the mission; to Christianize them; to teach them artisan skills, farming, and animal husbandry; to change them, in short, from savages into "people of reason," and having done this, to move forward with the advancing frontier. Their onetime charges, sufficiently Christianized to depend on a secular priest, responsible enough to be entrusted with a private share in the old collective lands, were also ready, the theory went, to assume the role of yeoman farmers and, with no looking back, to assimilate themselves into the culture of the Spanish settlers by this time living among them. But in practice the pattern departed from theory in a number of respects, and the missions of the north, once founded, tended to persist.

By 1650 the system extended up the Yaqui Valley to the headwaters of the Sonora, San Miguel, and Bavispe Rivers—nearly to the present International Boundary. For the next one-half century few new establishments came into existence. Then, beginning about 1695 a fresh surge of activity, stemming from the remarkable energies of Francisco Eusebio Kino, sent a second wave of missions spilling out across the homeland of the Upper Pimas.

This surge marked the northernmost expansion of New Spain in the inland west, and after Kino's death in 1711 the sporadic attempts to convert the Indians of the San Pedro, Gila, and lower Colorado valleys met with no permanent success. In fact, the Jesuits, engaged in an unequal struggle with the settlers, the Apaches and finally, the crown itself, were hard put merely to maintain what they had won.

The colonization of Sonora. Not the least of the factors that made possible the rapid reduction of the native peoples by a mere handful of missionaries was the initial absence of white settlers from the region. On the coastal plain south of the Yaqui River a few ranches had been founded by 1600 and in the mountains, a few mines. By and large, however, the main stream of colonization came a generation later, when in 1637 Pedro de Perea received viceregal permission to found a colony on the San Miguel River at Tuape, several hundred miles in advance of the other Spanish settlements, and near the country that the Jesuits were then proselytizing. Although Perea, himself, soon fell from favor, the colony became a nucleus for rapid secular expansion (Mange 1926, Alegre 1956–60: III, 11–21).

By the time Kino undertook the reduction of Pimería Alta, gold and silver diggings dotted the uplands the length and breadth of the province, bringing in their wake ranches in the grasslands, Spanish and mestizo farms in the valleys. As the native population dwindled, European settlers encroached on the Indian villages, and in many places jostled the missions themselves. Increasing friction developed between the two cultures, bringing an increasing demand for secularization of the mission centers and for distribution of their lands. These the Jesuits resisted for two reasons. On a static frontier there were no new areas for them to move to. Moreover, they knew that the acculturation of the Indians required longer than the ten years allotted to it by the crown, and that to leave their charges merely meant abandoning them to the mercies of a society that had few (Dunne 1957: 87–88).

The quarrels of the eighteenth century. The period after 1711 became increasingly a time of conflict for Sonora, involving a many-sided struggle among missionary, colonist and Indian in which, by and large, the civil administration took the part of the settlers. Indian revolt constituted the most obvious manifestation of the conflict. In 1740 a general conflagration spread northward out of the Yaqui and Mayo villages. In 1751 the Upper Pimas rebelled (Ewing 1934). The Seris, implacable enemies of all Europeans after the abduction and enslavement of their women in midcentury, harassed the settlements from havens on Tiburón Island, along the arid reaches of the coast of the Gulf of California and in the Cerro Prieto. Presidio after presidio came into existence as the Spanish king attempted to control the unrest and to combat yet another enemy, who appeared from the north, the Apaches.

The Apaches. It is impossible to follow the movements of these seminomadic bands with any degree of certainty. They seem to have moved southward at about the time of the Pueblo Revolt in New Mexico in 1680, and certainly by 1692, when the Presidio of Fronteras was founded, they were raiding into New Spain and threatening the Sobaípuris of the San Pedro River.[17]

As the opulence of the colony increased, so did the tempo of the Apache raids. Missions, towns, and mines fell under attack, but the savages struck principally at the growing herds of livestock. Horses and cattle, driven through the thin defenses of the Spaniards, were consumed at leisure in strongholds to which troops rarely dared penetrate, and then only in such force that they could be evaded.

Before these hit-and-run tactics the ill-armed, ill-fed, ill-trained garrisons of New Spain, participating in the general decline of Spanish might everywhere, were virtually impotent (Chapman 1916: 65–66, Thomas 1941: 10–12).

Eighteenth century accounts are replete with references to the destruction and depopulation attendant upon the Apache raids. Nentuig (1951: 136–37) wrote that, "Sonora is on the verge of destruction, for we have seen . . . scarcely ten places inhabited by Spaniards, while there are more than eighty ranches and farms destroyed by the enemies." Pfefferkorn (1949: 43–45) stated that:

At present . . . Sonora is not a shadow of what it was, and there remains for it only the sad remembrance of its former prosperity. . . . These savages have for many years raged terribly in Sonora, have cruelly murdered or carried off into captivity a large number of Spaniards as well as converted Indians, have stolen an indescribable number of horses, mules, and cattle, and have committed other like devastations. Because of this, there has occurred a gradual exodus from Sonora of many and indeed, of the most well-to-do Spaniards, who have sought another abode where their lives and properties would be safe. . . . No one can engage in agriculture in outlying areas without risking his life. . . . These same savages have also almost completely ruined stock raising.

A certain amount of sophistry is necessary to compare, as Nentuig does, inhabited towns to abandoned ranches and farms; there are, moreover, strong overtones of frontier paranoia about the whole Apache menace. A matter-of-fact survey of his flock, about 1763, by the Bishop of Durango makes it clear that there were considerably more than "ten places inhabited by Spaniards," in Sonora. Indeed he enumerates 52 of them, 5 with a population in excess of 1000 (Tamarón 1937: 237–51, 279–86, 303–11).[18] By his account at least 17,000 *gente de razón* inhabited the province at the time Nentuig and Pfefferkorn were describing its near desolation, and to have achieved that level of population the province must have undergone continuous growth. Although the Apache-Seri warfare may have resulted in widespread murder and destruction, it does not seem to have interfered with the fundamental processes of colonial development.[19]

The reforms of the enlightenment. Internal dissension in Sonora and attack from without were only two of the problems besetting the sprawling panorama that made up the Spanish Empire. One and one-half centuries of centralized misgovernment, a mismanaged and moribund economy, a series of disastrous wars commencing with the defeat of the first Spanish armada and culminating in a debacle that saw Spain torn apart by the dynastic ambitions of Europe—all combined, by 1759, to reduce the proud land of the conquistadors to the status of a second-rate power. Over her decline there had presided a succession of kings ranging from merely incompetent to, in the case of the unhappy Charles II, literally imbecilic. In the person of Charles III, who came to the throne in 1759, she found a ruler the likes of whom she had not seen in 150 years, and Charles addressed himself with vigor to the whole complex of domestic and colonial ills.

In connection with the troubles on the northern frontier he dispatched José de Gálvez to the New World on a tour of inspection (Priestley 1916). Gálvez' visit resulted in the creation of the Interior Provinces, a sprawling jurisdiction partially independent of the viceroy and governed from Arizpe by its own commandant-general; the shuffling of the presidial line and its temporary extension northward to the junction of the San Pedro River and Babocomari Creek.[20] Able administrators like Teodoro de Croix and Juan Bautista de Anza reflected the competence of their superiors, and under Gálvez' direction moved toward the rejuvenation of the north.[21] A policy that coupled force with feeding brought the Apaches under control; for the next fifty years Sonora enjoyed a measure of peace.

But a second action by Charles offset, for the northwest, much of the good of the first, and is difficult to understand except in the context of the domestic situation in Spain. Aimed at ameliorating that, it brought temporary chaos to the northern frontier. It effectively ended the bickering among the Europeans, but at the same time it destroyed the equilibrium between Indian and white. In 1767 Charles expelled the Jesuits from his domains.

Before the Franciscans arrived to take charge, the chain of missions fell apart, and the position of the Indians suffered irreparable damage (Thomas 1933: 306). Despite the efforts of the Franciscans to restore the system to its old vigor, the waning years of the eighteenth century witnessed the secularization of many centers, the further encroachment by white settlers, and the rapid disappearance of the native peoples into the semi-Europeanized, hybrid mass that for sheer bulk, was coming to dominate New Spain—the mestizo.[22]

With the death of Charles III in 1788, Spain's spasm of vigorous and effective government ended. Under Charles IV she resumed her old ways and so did her frontiers. But time was running out everywhere for the Bourbon way. France deposed her branch with a revolution; the revolution gave way to the Directory, and it to Napoleon. With him the Spanish Bourbons fell upon hard days, and on him the Spanish Empire shattered. After Napoleon one must speak not of Spain, but of Mexico in the Sonoran Desert.

The ecological impact of the Spaniards. During the two centuries of their rule in the desert region, the Spaniards came to overshadow the Indian and to exercise relative to him an ecological importance commensurate with their political dominance.

Their place in the regional ecology has two aspects. First they altered the Indian's relation to his environment. Secondly they created a role of their own.

The first of these may be summarized by saying that the number of aborigines declined in the wake of the European invasion, and that the geographical distribution of the survivors was transformed as a result of the Jesuit policy of concentrating their wards in mission centers. Once settled, the decimated Indian population came to rely more heavily than before on cultivated foods, although game and wild plants still made up an important part of their diet.

In one respect then, the Spaniards turned back the clock. Food gathering and hunting declined in importance to their level at some indeterminate time prior to 1600. It is difficult to avoid the conclusion that if there were one-third as many Indians in 1800 and they relied only one-half as much on wild food, their impact on the wild life would be, on the average, only one-sixth as great as it had been in 1600 when the Europeans arrived.

The decrease, however, was not uniformly distributed. Around the populous mission centers in the valleys, Indian disturbance may have been as great as before, or greater. In other parts of the desert, beyond the food-gathering range of the missions, it may have dropped to inappreciable levels.

One can envision, after 1600, a slow recovery of the native vegetation over most of the region from whatever Indian pressures it had been subject to; an intensification of the disturbance in the valleys where appreciable Indian congregations developed. But to a considerable extent, any tendency toward overall recovery was swamped by the impact of the new set of factors associated with the Spaniards.

Spanish cattle raising. In the first place, they introduced domestic grazing animals—cattle, sheep, and goats. On the Central Plateau in the first half-century after the conquest the native population fell to one-half its former level; the number of sheep grew to about 2,000,000; cattle to about 200,000. The next fifty years saw the human population reduced again by about one-half, the number of sheep increased to 8,000,000 and cattle to 1,000,000 (Simpson 1952: frontispiece).

Similar figures for the livestock population of the desert northwest are lacking: our present knowledge is barely sufficient to sketch the outlines of the expansion. By the end of the sixteenth century, "the ranching frontier of New Spain pushed in a rough arc northward from Culiacán to a crest on the southern tributaries of the Conchos and then turned south toward Monterrey" (Morrisey 1951). By 1610 there were ranches near the southern fringe of the Sonoran Desert between the Sinaloa and Fuerte Rivers (Ewing 1934: 18). In 1678 the Visitador Don Gabriel de Isturiz inspected ranches at Huépac and in the valleys of Bacanuchi and Cedros (Bannon 1955: 137–39). By 1694, according to Herbert Eugene Bolton (1948: map), 100,000 cattle ranged the grasslands around the headwaters of the San Pedro and Bavispe Rivers.

The spread of cattle through the mission chain supplemented the secular diffusion; as each new establishment was founded, it received a herd from the older missions. Kino, by himself, introduced cattle over much of Pimería Alta, and his *Memoir* is replete with references to gifts of livestock. His home herd at Dolores in 1701 provided 1400 head for Baja California and San Xavier del Bac; in 1702, 700 more. In 1703 there were yet another 3500 available for distribution (Bolton 1948: I, 58; I, 357).

It seems safe to assume that, by 1700, cattle ranged over virtually all of the inhabited parts of the desert region on the mainland of present-day Mexico, except the arid wastes along the lower Colorado Valley and the coast of the Gulf of California. By the mid-eighteenth century, when Pfefferkorn wrote, individual secular herds of four or five thousand head were commonplace (Pfefferkorn 1949: 43, 45, 91, 94–95, 98, 285).

The extent to which Apache raiding acted as a deterrent to cattle-raising has probably, like the Apache menace in general, been somewhat exaggerated (*e.g.* Pfefferkorn 1949: 45, 98). In the first place the Apaches preferred horsemeat (Bancroft 1883: I, 490). In the second place the bands were small; even if they had confined themselves to an exclusive diet of beef the amount consumed would probably have remained below the replacement capacity of the herds. In the third place, the Apaches followed a hit-and-run technique; large cattle drives could not have been made swiftly and without detection, even through the inadequate defenses of the northern frontier.

A report by Governor Barri of Nueva Vizcaya provides some concrete information about the magnitude of the losses. For that immense province, larger than Sonora, with more cattle, and subject to raids from Apache and Comanche alike, Indians stole 68,000 cattle, horses, and mules in the five years between 1771 and 1776 (Thomas 1941: 31). An annual loss of 13,000—perhaps 1 or 2 per cent—is

hardly critical, although the losses to individual cattlemen may have been severe. If, as Barri also reports, the 116 haciendas and ranches that were abandoned in the same five-year period were indeed abandoned, and were abandoned because of Indian raids, fear may have been a more important factor in limiting the number of livestock than the physical theft of the herds.

The presence of cattle in large numbers at this early date raises a number of ecological questions, since overgrazing has been widely blamed for the spread of mesquite and for the arroyo cutting in southeastern Arizona after 1880, when cattle became numerous in that area. Is there any evidence that comparable processes began two centuries earlier in Sonora?

About the prevalence of mesquite in the grasslands there is little evidence one way or the other; certainly it was already abundant in the lower desert valleys. As for arroyo cutting, one can be reasonably sure that it did *not* commence in 1700, but made its appearance along the Sonoran streams about the same time as in Arizona—nearly two centuries later.

In 1931 Sauer and Brand reported that local tradition attributed channeling in the parts of the Altar Valley around Santa Teresa and Atil, Pfefferkorn's old mission, to the earthquake of 1887 and to the floods of 1905 and 1927 (Sauer and Brand 1931: 98). At the time of their visit the Dolores Valley was still untrenched around Kino's home mission, but twelve miles away at Ojo de Agua, "a flood in 1914 ripped out a good part of the valley floor. . . . It is said that prior to that time there was no intrenched channel, but that the water flowed off through ciénegas and open pools" (Sauer and Brand 1931: 85).

They called attention to recent arroyo cutting in several other localities, and concluded (pp. 122–24) that:

If one can take the general testimony of old residents, the past forty to fifty years are responsible for the whole of this trenching. In that time the arroyo beds, in particular those of the normally dry basins, have been sunk by as much as twenty to thirty feet. These sandy washes usually are growing steadily wider and deeper and are extending headward. . . . Some of the arroyo floors have grown a quarter of a mile wide and even wider, blotting out what was the moistest and normally most desirable land until the present generation, and replacing farm land by a sand waste. In these ways a serious and continuing degeneration of the basin floors has set in, which has much reduced, and in some cases has nearly eliminated, the agricultural potentialities . . . of historically productive localities, such as the missions of Santa Teresa and Búsani.

The chronology has some important implications: if arroyo cutting began at about the same time in both southeastern Arizona and northern Sonora, although extensive stock-raising commenced in Sonora two centuries earlier, what basis is there for assuming that cattle-raising had anything to do with the process?

Qualitatively expressed, the question answers itself, and the answer lends no support whatever to the chain of cause and effect propounded by writers who have considered only the events of the 1880's, without looking any farther back in history than that period, or any farther south than the streams of southeastern Arizona.

However, when the question is properly expressed, with the necessary quantitative factors in place, the answer is not obvious; one cannot be sure of it until more is known about the growth of Mexican ranching. What needs to be determined is not whether the number of cattle per unit area in 1890 in southeastern Arizona exceeded the density in northern Sonora in 1700; this comparison introduces too many intangible factors relating to differences in soil, flora, and climate between the two regions. What needs to be determined rather is whether there were as many cattle in Sonora in 1700 as in 1890.

If there were not, the hypothesis is still tenable that large-scale stock-raising led to arroyo cutting. If there were, the hypothesis has to be abandoned. Or at any rate it has to be modified to the extent that it considers "overgrazing" to be dependent as much on range conditions as on the number of cattle, as much on the grass cover produced by climate as on the amount removed by grazing. It is not inconceivable that the ranges of Sonora were as crowded in 1700 as they were two hundred years later, but that because of more favorable conditions of rainfall and temperature they were able to support the earlier herds. In a case like this (which is far from unlikely), either-or reasoning breaks down. What "caused" the events of the 1880's? Not large-scale cattle-raising certainly, and perhaps not climate; probably the combined impact of the two.

Spanish mines and ecology. Mining constitutes a second point at which Spanish activity impinged on the ecology of the desert region. On the alluvial plains the effect may have been slight; the impact on localities in the uplands, where the principal ore deposits lay, may be judged from studies of similar areas elsewhere in Mexico.

West (1949: 39–46) has concluded that the scrub grassland around the mining town of Parral used to be oak woodland, and that the transition from one to the other occurred during colonial times. Making charcoal, timbering shafts, smelting the ore, and keeping the inhabitants warm made such heavy demands on the local fuel resources that oak was wiped out. At the same time, overgrazing by pack animals reduced the grass cover on the ejido lands and induced erosion there.

Describing another area about 1604, Mota y Escobar says: "At the time of the discovery of the mines the hills and canyons around Zacatecas were covered with extensive stands of trees, all of which have been cut down for use in the smelters. Only a few wild yucca remain. Wood is very dear in this town, for it is brought in carts from a distance of eight to ten leagues" (West 1949: 116, n. 166).

West describes other instances of depauperate vegetation near Parral, and Sauer (1956: 63) states that: "Quite commonly the old mining reales of North and South America are surrounded by a broad zone of reduced and impoverished vegetation."

There is little room for doubt that mining, where it was carried on extensively, did exercise an important effect on the plant life, and was a potent factor in shaping parts of the landscape as it had come to look by 1800.

The Spaniards in retrospect. As the century drew toward its close, and the time grew near for their departure, the Spaniards could look back on two hundred years in which they had profoundly altered the life of the northwestern Indian and had, to a more modest degree, touched the other life of the desert region. One of their activities, stock-raising, probably was as important as any cultural factor has ever been. A second, mining, although less far-reaching, resulted in some changes that were locally important. In both of these activities the Spaniards anticipated the Americans. The ecological effects of the latter people may not, in light of what their predecessors did, have been so unique as they are usually portrayed.

THE MEXICANS

With Hidalgo's *Grito de Dolores* in 1810 the Mexican movement for independence formally commenced; after 1821 when Agustín Iturbide elevated himself to an unsteady throne, a new nation appears on the scene.

The history of the generation during which Mexico ruled the Sonoran Desert by herself may be traced very briefly. To a great extent the ecological trends inaugurated by Spain continued, and the amount of political history that is pertinent is slight.

At national level the period was characterized by the *caudillismo* of Antonio López de Santa Anna. In the Northwest the years until 1830 were consumed in a quarrel over the question of state boundaries. In 1823 Sonora withdrew from the jurisdiction of the commandant-general at Chihuahua and established her own government at Ures. In 1825 Occidente, a state comprising both Sinaloa and Sonora, came into existence. From 1825 until 1830 quarrels among Ures, Arizpe, Culiacán, Fuerte, and Alamos over which should be the capital kept Occidente in turmoil. In 1831 Sonora again constituted herself as a separate entity, and thereafter until the 1850's the figure of Manuel Gándara moved in and out of state affairs in a way reminiscent of his mentor, Santa Anna, on the national scene (Bancroft 1889: II, 628–48).

At a time when she could ill afford it, Sonora found herself plagued by other misfortunes. The "loyal" Opatas rebelled in 1820, the Yaquis in 1825, again in 1832, and at intervals thereafter until the twentieth century. Federal edicts expelling the Spaniards from the Republic and secularizing the remaining missions gave rise to further dislocations (Bancroft 1889: II, 633–52). Amid the prevalent civil and political disorder the state could take comfort only in the relative peace she enjoyed at the hands of the Apaches.[23]

By 1800 the policies initiated by José de Gálvez had virtually put a stop to raiding, and the truce apparently continued through the otherwise troubled period when Mexico was asserting her independence from Spain (Stevens 1964). From 1810 until 1831, when the Indians struck in force once more,[24] the Sonoran cattle industry migrated northward into Arizona (Bancroft 1962: 399–407, Bancroft 1889: II, 653 ff.). Insofar as this study is concerned, that expansion is the principal cultural fact bearing on the ecology of the region during the Mexican period.

The northward expansion of stock-raising. To the Santa Cruz and the San Pedro Valleys the activities of white men were hardly new. Along the former, the more active of the two valleys, three missions, Guébabi, Tumacacori (Plate 40), and San Xavier, had been erected by the time of Mexican independence, all dating their origins, if not their buildings, from Kino's day. In the wake of the Pima uprising of 1751 a presidio at Tubac had come into existence. Occasional agricultural activity had taken place along the river as far north as Tucson.

With these exceptions no important, permanent Spanish settlements were founded north of the present International Boundary until the nineteenth century. Then, with the lessening tempo of the Apache raids, large-scale stock-raising began. In 1806–7 Spain granted title to the Indians for large tracts of land around Tumacacori and Calabasas to take care of their livestock (U. S. Congress 1880: 18–24).

In the years between 1820 and 1822 substantial grants were made to Spanish and Mexican cattlemen along the northward flowing part of the Santa Cruz at Canoa; along the southward flowing part in the San Rafael Valley; on Sonoita Creek; and around San Bernardino, southeast of present-day Douglas.[25]

The two years following 1826 witnessed still others: the Buena Vista grant along the Santa Cruz River near the present International Boundary (Plates 22–24); the Boquillas and San Rafael del Valle grants on the San Pedro River (Plates 50–51, 46–47); a grant along Babocomari Creek (Plate 57).[26]

These Spanish and Mexican titles later gave rise to extended litigation in the United States courts. Here the lands are troublesome because of the difficulty in deciding how intensively they were used. The point is primarily an ecological one: if large herds ranged over them during the 1820's and 1830's, why were arroyo cutting and mesquite invasion delayed until half a century later? The historical evidence fixes some rough limits for the number of cattle involved: at one end of the scale, the livestock population certainly did not approach the levels of the 1880's. There were only a few ranches, probably not more than those already enumerated,[27] and these were confined to the valleys. Furthermore, the period of peace did not last long enough for big herds to develop.

By 1831 Apache activity brought a halt to the attempts to develop a grant at Tres Alamos on the San Pedro River below modern Benson.[28] In 1833 and again in 1836 the Herreras family was driven from its lands on Sonoita Creek (Mattison 1946: 300). None of the grants later upheld by the U. S. courts was made after 1831, and one can infer that after that time conditions for settlement again became too adverse.

By 1846 all of the estates were abandoned. By the time of the earliest American accounts—on the heels of the Mexican War—the area north of the present boundary was uninhabited except for a few ranches along the Santa Cruz River that saw sporadic occupation as conditions permitted.

Although early American accounts help the historian circumstantially to assert that because of its early abandonment, cattle-raising did not attain the proportions of later years, at the same time they define a minimum level of activity that must have taken place in order to account for the impressive remains of the industry.

Philip St. George Cooke, whose Mormon Battalion passed through San Bernardino in 1846, described that place as follows:

This old ranch was abandoned, I suppose, on account of Indian depredations. The owner, Señor Elias, of Arizpe, is said to have been proprietor of above two hundred miles square extending to the Gila, and eighty thousand cattle. Several rooms of the adobe houses are still nearly habitable. They were very extensive, and the quadrangle of about one hundred and fifty yards still has two regular bastions in good preservation. In front and joining was an enclosure equally large, but it is now in ruins (Bieber 1938: 132–33).[29]

John Russell Bartlett, lost along the Mexican Boundary he was supposed to be surveying, stumbled onto the old hacienda of San José de Sonoita, "A cluster of deserted adobe buildings . . . which seemed to have been abandoned many years before, as much of its adobe walls was washed away." [30]

En route back to the San Pedro River he wandered into the headquarters of the old Babocomari ranch:

This hacienda, as I afterwards learned, was one of the largest cattle establishments in the State of Sonora. The cattle roamed along the entire length of the valley; and at the time it was abandoned, there were not less than forty thousand head of them, besides a large number of horses and mules. The same cause which led to the abandonment of so many other ranches, haciendas, and villages, in the State, had been the ruin of this. The Apaches encroached upon them, drove off their animals and murdered the herdsmen; when the owners, to save the rest, drove them further into the interior, and left the place.[31]

At Calabasas he inspected the "ruins of a large rancho" (Bartlett 1854: II, 307), and at San Lázaro, "A large deserted hacienda" (Bartlett 1854: II, 311–12), where three years earlier H. M. T. Powell had noted "vast piles of bones of cattle. . . . The corral which was burnt up, could have easily held a 1000 head of cattle" (Powell 1931: 136, Evans 1945: 148).

Tubac, Tumacacori, Guébabi, and other locations are also described in the accounts; at times they were deserted, at times, not, depending evidently on the current extent of Apache raiding.[32]

Not only the ruins of the haciendas, but the size of the remnants of the old herds, go to establish the extensive nature of the Mexican undertakings: "Many of the cattle . . . remained and spread themselves over the hills and valleys near [Babocomari]; from these, numerous herds have sprung, which now range along the entire length of the San Pedro and its tributaries" (Bartlett 1854: I, 396–97).

In 1846 Cooke, encamped at a watering place a few miles east of Agua Prieta Creek, wrote that: "the wild cattle are very numerous. Three were killed today on the road and several others by officers. Around this spring is a perfect cattle yard in appearance, and, I suppose, I myself have seen fifty. . . . It is thought that as many as five thousand cattle water at this spring" (Bieber 1938: 132).

All the way along the deserted San Pedro Valley the battalion dined on beef; at the junction of the river with Babocomari Creek, they fought their first engagement of the Mexican War—with a herd of wild bulls: "One of them knocked down and run over Sergeant Albert Smith, bruising him severely. . . .

One of them tossed Amos Cox of Company D into the air, and knocked down a span of mules, goring one of them till his entrails hung out, which soon died." [33] "Lieutenant Stoneman was accidentally wounded in the thumb" (Bieber 1938: 143).

The frequency with which the journals mention them between 1846 and 1854 leaves no room for doubt about the abundance of the wild herds (Durivage 1937: 205; Cox 1925: 140–42; Evans 1945: 144–45; Clarke 1852: 78; Powell 1931: 126–28). With a rough maximum and minimum fixed—that the number of Mexican cattle on the grants was substantial, but did not approach the livestock population there in the 1880's, one can guess that the actual number lay between 50,000 and 150,000.

Ecologically, then, cattle-raising was extensive enough to result in a thorough dissemination of mesquite seeds; insufficient to reduce the grasses to the impoverished condition that they knew by 1890; sufficient to constitute the severest cultural stress yet imposed on the region. Certainly the assault of 100,000 head of livestock on the vegetation dwarfed any food-gathering activities of the earlier Indian population, numbering some 10–15,000.

Whether as a result of the transient invasion in the 1820's and 1830's any important vegetation changes took place, one cannot be sure, but the answer is probably that they did not. There is nothing in the early American accounts to indicate that the country looked any different from the picture of it that emerges dimly from the early Spanish records. The two agree that grass was plentiful; that the landscape was open; that mesquite, though widespread, had nothing like its present density. The streams at both times were marshy, open, and largely unchanneled. Beyond this, the image cannot be resolved more closely.

The transition to the American period. By 1846, the native cultures, except for the Apaches', had seen their best day; the Spaniard had come and gone; the Mexican remained, but in the northern part of the desert his influence had waned.

To the east the Manifest Destiny of the United States was carrying that nation vigorously westward. In 1846 the old and the new cultures met, and by 1848 everything north of the Gila River belonged to the United States. In 1853–54 Santa Anna, giving his final performance before the Mexicans, sold their land between the Gila and the present International Boundary.

Thereafter the desert region knew two political masters, and its ecology was subject along an artificial but, nevertheless, important line to two distinct and, in many ways, divergent sets of cultural factors.

III

THE INFLUENCE OF MAN:
THE ANGLO-AMERICANS

INTRODUCTION

The earliest American accounts of the desert region date from shortly after the Mexican War for Independence when the mountain men—the early fur hunters—taking pelts where they found them, trapped beaver along the Gila and its tributaries.[1]

As the western migration swelled, and well before title to the lands south of the Gila passed to the United States, travel across the region became commonplace. For the decade prior to the Gadsden Purchase, there exist a score of journals describing the country, most of them the offspring of troop movements during the Mexican War or the rush of the forty-niners across the continent on their way to the gold fields of California. Taken as a whole, the early accounts present a rich and detailed picture of the natural conditions that prevailed in the forties and fifties.

The landscape that they describe lasted for another generation, and the cultural factors that had helped to shape it changed very little in the 1850's. Some settlement took place immediately after the Gadsden Purchase; a few military posts came into existence (Eaton 1933); a few mines opened up in the hills. By and large, however, the years until the Civil War constituted for the region's ecology merely an extension of the Mexican period.

CONDITIONS UNTIL THE CIVIL WAR

From the early journals there emerges a picture of the country that in many respects contrasts sharply with present conditions. In particular the appearance of the valleys and the grasslands has changed (Hastings 1959).

The valleys. Today the streams in the Arizona part of the region—the San Pedro, the Santa Cruz, and their tributaries—maintain highly irregular regimes. Except in their upper reaches or along the parts of their course where rock formations force water to the surface, the beds much of the time are dry, sandy wastes that support little, if any, vegetation. Five to thirty feet above the channel and set apart from it by abrupt vertical banks, one typically finds a bosque dominated by mesquite. During the summer rainy season, flash floods render the streams impassable. At such times the channel is filled bank to bank with a raging, muddy torrent that carves new incisions into the flood plain and sharpens the edges of the old.

Before the Civil War these same streams wound sluggishly along for much of their course through grass-choked valleys dotted with cienegas and pools. In spite of the onslaught by the mountain men, beaver dams were still numerous, and as late as 1882 a settler on the San Pedro could report that:

Our ditch was just above a beaver dam and if the water was low and we tried to irrigate at night the beaver would stop up our ditch so that the water would run into their dam.[2]

Of the San Pedro, Cooke noted a fact equally improbable from the perspective of the twentieth century: "Fish are abundant in this pretty stream. Salmon trout are caught by the men in great numbers; I have seen them eighteen inches long." [3]

Although they flowed more regularly than they do today, the rivers nevertheless did go dry at times or along particular stretches:

The San pedro river as they Call it—is a stream one foot deep six feet wide and runs a mile and half an hour and in ten minutes fishing we Could Catch as many fish as we Could use and about every 5 miles is a beaver dam this is a great Country for them—and we have went to the river and watterd and it was running fine and a half mile below the bed of the river would be as dry as the road—it sinks and rises again (Tevis 1954: 55).

Exceedingly to the surprise of every member of the expedition who had passed over this route in the months of March and April it was discovered after a march of a few miles that the waters of the San Pedro had entirely disappeared from the channel of the stream. . . . Where the present reporter took quantities of fine trout in March and April 1858 not a drop of water was to be seen [in September, 1858] (Itinerary 1858: 33).

Similarly, although there was far less channeling than at present, there was some, with the river's course defined in varying degrees. An early picture of the junction of the San Pedro River and Babocomari Creek appears as Plate 57. The creek seeped through a rank growth of marsh grass and can hardly be said to have a course. Near Tucson, at a place where its channel is now ten feet deep and seventy-five feet wide, the Santa Cruz in 1849 ran in an untrenched, but well contained stream better defined than the Babocomari's, and perhaps with multiple channels:

We encamped in a grassy bottom, much covered with saline efflorescence. The river has divided to a mere brook, the grassy banks of which are not more than 2 yards apart (Powell 1931: 145).

A distinct channel trench existed in places along the San Pedro and probably along the other streams:

The valley of the San Pedro River [near the mouth of Dragoon Wash] . . . was anything but luxuriant. It consists of a loam, which if irrigated might be productive, but as the banks are not less than eight or ten feet high, irrigation is impracticable. . . . The grass of the vicinity is miserably thin and poor, growing merely in tufts beneath the mesquit bushes which constitute the only shrubbery, and in some places attain a height of ten or twelve feet. . . . In order to cross the river, it was necessary to level the banks on both sides, and let the wagon down by hand.[4]

A few miles farther north Bandelier (1892: pt. 2, 478), writing just before the severe erosion of the 1890's, found "a cut with abrupt sides . . . 10 to 15 feet deep, and about 25 wide." In Plate 51, a picture of Charleston about the time of Bandelier's visit, a clear, small trench is visible.

About the presence of brush, there is similarly no simple generalization that will hold. It is apparent from many of the old photographs and accounts that parts of the valleys were more open than at present. Nevertheless brush and even bosques are frequently mentioned.

The valley of this river [the San Pedro near its junction with the Gila] is quite wide, and is covered with a dense growth of mezquite . . . cotton wood, and willow, through which it is hard to move without being unhorsed. The whole appearance gave great promise, but a near approach exhibited the San Pedro, an insignificant stream a few yards wide, and only a foot deep (Emory 1848: 75).

The bottom of the San Pedro is one mile broad [at the same site], and of the character of those on the Gila above, dusty dry soil, grown in places with cottonwoods and willow, in others with grass and again mesquite, chapparal, other places bare (Johnston 1848: 592).

The land on each side of the Pedro river bottom is a dense thicket of bramble bush, mostly muskeet, with which millions of acres are covered (Golder, Bailey, and Smith 1923: 193).

A description appearing in the documents pertaining to the Leach Wagon Road makes it clear that the contrast between mesquite forest and open valley was sometimes sharp:

A forest of heavy mesquite timber about one mile in width extends from the [San Pedro]. . . . Running nearly due north a road lies opened in March last through the Mezquite forest . . . for a distance of about three miles. . . . Leaving the forest it enters upon a tract of the bottom lands of the San Pedro . . . from one fourth of a mile to one mile in width. . . . The entire body of these lands were covered with a dense growth of sacaton grass . . . (Itinerary 1858: 32).

A second account by the builders of the wagon road describes the same part of the river:

The San Pedro, at the first point reached in the present road, has a width of about twelve (12) feet, and depth of twelve (12) inches, flowing between clay banks ten or twelve feet deep, but below it widens out, and from beaver dams and other obstructions overflows a large extent of bottom land, forming marshes, densely timbered with cottonwood and ash, thus forcing the road over and around the sides of the impinging spurs (U. S. Congress 1859: 87).

Collating the two descriptions yields the information that the open, grassy bottoms were marshy, and that at the point in the bosque where the road intercepted the river, a channel trench already existed. In the account by Bartlett already quoted, the same identification exists between mesquite land and a shallow, adjacent arroyo.

Given these clues, one can partially reconstruct the pattern of riparian vegetation. Along the parts of the river where there was no trench, the water table was high, the bottoms marshy, the soil waterlogged and too poorly aerated during at least part of the year to support anything but marsh vegetation dominated by grass.

Where there existed an arroyo, on the other hand, the bottom of the trench fixed the elevation of the water table. Between the surface of the water and the top of the bank there existed a layer sufficiently well drained to support mesquite and other plants whose roots require aeration and cannot tolerate waterlogging.[5]

This appears to be a reasonable explanation, but it may be oversimplified. An 1854 description of the river about two miles above the point where the Leach road reached it, warns of some pitfalls:

At the Tres Alamos [crossing] the stream is about fifteen inches deep and twelve feet wide, and flows with a rapid current over a light, sandy bed, about fifteen feet below its banks, which are nearly vertical. The water here is turbid, and not a stick of timber is seen to mark the meandering of its bed (Parke 1857: 25).

Here a treeless condition is equated with the presence of a channel trench. But this account, in turn, must be read in light of another (Eccleston 1950: 192–93), a few years earlier, of the same crossing:

This river, the San Pedro, is extremely boggy & has to be crossed by making a brush bridge. . . . I cannot agree with Colonel Cook, who calls this a beautiful little river, although where he crossed it, some 10 or 15 miles above, it may have presented more amiable qualities. Here it is lined with a poor growth of swamp willow & other brush, so that it cannot be seen till you come within a few feet of it; & then the bank is perpendicular, not affording an easy access to its waters, which, though not very clear, is good. The banks & bed are extremely boggy, & it is the worst place for cattle & horses we have yet been, being obliged to watch them very close.

In spite of the channel trench the banks were boggy. The identification between marshiness and treelessness is still valid. The picture is complex, but no more so than today, and no more contradictory. Many of the old cienegas—the distinctive, treeless marshes of the Southwest—drained by a channel trench through their middle, have disappeared into a sea of mesquite. Yet San Bernardino Cienega, located in northeastern Sonora, is still intact and brush-free in spite of the deeply incised channel of San Bernardino Creek which flows immediately to the east of it. Evidently a great deal depends—and depended—on the soil, how rapidly it erodes, whether water percolates through it at a rate in excess of the supply.

For the abundance of marshes along the valleys, there is extensive documentation:

The wild turkeys hatched in places where the marshes were, during those days, and they were protected by the grass that grew waist high. There were marshes along the Aqua Fria [a ranch on the Santa Cruz River], right there at Calabazes, and up at Buena Vista. . . . These marshes have all dried up (Allison n.d.).

This bottom down here just back of and south of A mountain [near Tucson] was marshy. We used to have a ranch in there when we first came here from Calabazas. There was a thirty-seven acre lake, just south of A mountain, in that bottom. This lake was made by Solomon Warner. There was marshy ground over there, and he put in a dyke and planted willows along the dyke to protect it (Allison n.d.).

From the court records of an early suit over water rights there can be established the existence of a large cienega extending along the San Pedro from about modern Benson to old Tres Alamos (Cochise County District Court 1889). The boggy, but treeless and trenched location already discussed lay at the foot of it. Powell (1931: 130–31) complains about the heavy going in the upper San Pedro Valley southeast of the Huachuca Mountains, and Bartlett discusses a "boggy plain" in the same area.[6]

Indirect evidence for the degree of marshiness appears in the high incidence of malaria, a disease almost unknown today. In 1868 so much existed at Camp Grant, located at the junction of Aravaipa Creek and the San Pedro River, that the army set up a temporary convalescence camp twenty-eight miles away (Report on Barracks 1870: 465–66). Similar conditions prevailed along Babocomari Creek, Sonoita Creek, and the Santa Cruz River (Bourke 1950: 8; *Arizona Daily Star,* July 21, 1880; *Arizona Weekly Star,* September 25, 1879; *Arizona Citizen,* August 29, 1874; Spring 1903; Rothrock 1875).

In summary, the valleys were wetter and more open than today, and relatively unchanneled. But the precise conditions varied from place to place and probably from time to time. As the tributary washes dumped greater or less amounts of debris, depending upon where heavy summer rains may have struck, the rivers had to transport varying loads of sediment at different points along their course. Channeling and filling, aggradation and degradation—all may have been going on simultaneously, in various stages of development along various parts of the stream. If this dynamic situation existed, one can be sure that the vegetation reflected it. At a given time there may have been mesquite invading, where a temporary trench had sliced through the old flood plain, draining it; mesquite dying where the plain was aggrading and marshes being developed. The old accounts present a picture that is neither homogeneous, nor static. By postulating a dynamic situation one can reconcile the variety of conditions that evidently existed.

The grasslands. After the river valleys, the grasslands of the desert region appear to have changed most. The overall tendency, as the plates will attest, has been toward shrubbiness and a less open, less expansive landscape.

On the bajada of the Sierra San José, which in places now supports a dense cover of spiny shrubs, one early group of emigrants went without breakfast because there was no wood for cooking (Powell 1931: 130). Another journal (Bartlett 1854: II, 325) relates that "the men picked up every fragment of wood or brush [they] passed." In the Sulphur Springs Valley, Parke (1857: 24) described the Willcox Playa as "bounded by smooth and grassy plains." At Croton Springs on the northwest side of the Playa, wood was "distant" (U. S. Congress 1859: 99), and "fire wood [had to be] dug from the ground, the roots of the Mesquite being the sole dependence" (Itinerary 1858: 30).

All of these areas are characterized today by shrubby vegetation, and a short distance from Croton Springs there is an abundant supply of wood, indeed. But at the same time, where there were mesquite roots in 1858, there must have been some mesquite. The situation was very far from being as black and white as modern interpretations sometimes make it.[7]

Along San Simon Creek, Parke (1857: 24) described a situation that may have been typical of large expanses of the desert grassland:

This plain presents the same features as the [Willcox] playa just described, as far as regards vegetation, without being absolutely bare, as that was; yet its growth is of that thorny, worthless, desert character. Fouquieria, larrea, yucca, palmetto, and agave, are the only growths on the slope of the plain, down to half a mile from the river, where the mesquite tree begins to appear, and the willow is found collected round some of the water holes in the bed of the stream.

It would be an exaggeration to think of the desert grassland as being uniformly like the midwestern prairies—open, rolling, and treeless. Parts were—and still are—but the chances are that most parts were not. Although the past century has seen a striking increase in the number of spiny shrubs and small trees, the invasion has been in the nature of an increase in the density of the species that were already present, not (except for introduced exotics like Russian thistle) an extension of the range of new plants into areas that they formerly did not occupy.

Fire in the grasslands. Whether the suppression of fire brought about the shrub invasion is a matter on which the early American accounts, extensive as they are, should shed some light.

The point has already been made that the control of shrubs by burning is a quantitative, not a qualitative matter. That the Apaches used fire as a hunting tool proves nothing about its effect on the plant life. A rough analogy exists to food-gathering activities. That the Seris ate saguaro fruits, a qualitative fact, in no way establishes that the habit was important in determining the plant's distribution. Neither was the presence of fire, *per se,* important. The impact depended entirely on how frequently burning occurred and how extensively.

Ecologists and geographers have made wide use of historical evidence in studying the occurrence of prairie fires in the Midwest and Texas, to ascertain if they were frequent or extensive enough to maintain the grasslands in a shrub-free condition.[8] The evidence, although impressive, cannot be transferred to the Sonoran Desert region where the vegetation, the topography, the climate, and the habits of the native peoples were different.

But what can be done for one place can be done for the other. If early accounts of the Midwest do, in fact, demonstrate the existence of a fire climax, there is no reason why the abundant historical records of the desert grassland of northern Sonora and southern Arizona should not, if it was true, show the same thing. Furthermore, they should show it in a roughly quantitative fashion.

Humphrey (1958: 243) has observed that: "Invading shrubs . . . will normally not produce seed for several years. Fires occurring at intervals more frequent than this . . . would continue to keep them suppressed." This criterion seems acceptable, and although the exact interval would vary from species to species, one supposes that recurrent fires every ten years might suffice.

On this basis about one-tenth of the grasslands would have to be fired annually. For an individual year the figure might be higher or lower, but over several decades the irregularities would average out. However, at one point in time, during even an average year, something less than 10 per cent of the grassland might give the appearance of having been burned. An area that had suffered fire in spring might enjoy a vigorous new growth in summer and bear no marks of burning at all by fall. To allow for biseasonal vegetative activity, the 10 per cent can be reduced to 5.

In traversing the desert grassland, then, travelers in the forties or fifties should, on the average, have observed one mile of burned grass for every twenty traversed. The extent to which they may have recorded the observations is another question. No one wrote down all the minutiae of a long trip; on the other hand a tendency to glide over the obvious or uninteresting is offset in this case by the sensitivity of most emigrants, because of their mounts or their teams, to grass conditions.

In any event the probability of getting an accurate idea of the extent of burning can be maximized by taking only the journals kept on a day-to-day basis, not those written in retrospect. It can be further maximized by choosing, wherever possible, expeditions for which several accounts exist; what one person forgot to record, another may have remembered.

A list of twenty-two travel journals meeting these minimum standards and spanning the early period of American travel through the desert grasslands has been assembled and appears in Table 1.[9] For many of the expeditions there is multiple coverage: for the Mormon Battalion, five journals; for the Army of the West, three; for the Leach Wagon Road, two and several fragments; for the first border survey, two; the journals of Evans and Cox belong together; those of Clark and Durivage. Altogether then, there are twenty-two separate accounts dealing with some twelve expeditions along various routes across the grassland during the period from 1846 to 1858. The number of traverses can be reduced to ten by lumping together three of the expeditions that passed over the same trail within a few months of each other (Evans-Cox, Powell, Aldrich), thereby merely sampling the same area three times. Altogether, then, there are ten traverses of about one hundred miles each, cover-

Table 1
Journals Examined for Evidence of Grass Fires in the Desert Grassland

Journal	Date	Route traversed
Cooke Standage Bliss Bigler Keysor (fragment)	Dec. 1846	San Bernardino—San Pedro River—Tucson
Couts	Oct. 1848	
Clarke Durivage	May 1849	
Evans Cox	Aug. 1849	San Bernardino—Santa Cruz—Tucson
Powell	Sept.–Oct. '49	
Aldrich	Oct. 1849	
Bell	Sept. 1854	
Emory Johnston Griffin	Nov.–Dec. '46	Gila River
Chamberlain	July 1849	
Eccleston	Oct.–Nov. '49	Stein's Peak—Railroad Pass—Tres Alamos—Tucson
Hutton Leach "Itinerary"	Apr.–Sept. '58	Stein's Peak—Railroad Pass—Tres Alamos—San Pedro River—Gila River
Bartlett Graham	1851–52	Throughout the border area from the Gila to Santa Cruz

ing a total of perhaps one thousand miles of travel through the desert grasslands. On the basis of the earlier calculations, one would expect the journalists to encounter about fifty miles of burned country.

For the parts of the journals covering travel between the Arizona–New Mexico border and the Santa Cruz River, all of the items pertaining to fire have been extracted, whether they deal with the grassland or not. They are as follows:

[The Army of the West, October 25, 1846, about 25 miles east of Safford, on the Gila River.] We passed today the ruins of two more villages similar to those of yesterday. . . . Except these ruins, of which not one stone remained upon another, no marks of human hands or footstep have been visible for many days, until today we came upon a place where there had been an extensive fire. Following the course of this fire, as it bared the ground of the shrubbery, and exposed the soil, etc., to view, I found what was to us a very great vegetable curiosity, a [barrel] cactus (Emory 1848: 243).

[Hays-Nugent emigrant party, October 26, 1849, descending Apache Pass toward the Willcox Playa.] There was a good deal of debate about what most supposed to be a large lake ahead; others affirmed it was sand, & some a fog. For myself, I was certain it was water. . . . Also another cause of debate was a fire we saw, and the question was, how far was it off (Eccleston 1950: 187).

[The Hays-Nugent emigrant party, November 1, 1849, near Cienega Creek about twenty-five miles southeast of Tucson.] Just about dinner time some of the Missourians somehow or other set fire to the prairie, which soon spread with the rapidity of lightening. It commenced on the opposite side of the branch about ¼ mile off. Our oxen were on the other side also, & our 1st efforts were directed towards them, & we ran as never men ran before. The grass, or rather cane, was some 6 ft. high, & where I crossed it it was extremely warm from the fire. It crossed the branch [Cienega Creek] & we now had to look after our waggons, etc. . . . The wind was blowing high and towards us. It did not go out till some time after dark (Eccleston 1950: 195).

[Bartlett, June 13, 1851, Guadalupe Pass.] Starting fresh, we hastened over the plateau, and soon entered the cañon in the Guadalupe Pass, which has so charmed us by its luxuriant vegetation and picturesque beauty. But what a change had taken place! A fire had passed over it, destroying all the grass and shrubbery, and turning the green leaves of the sycamores into brown and yellow. The surface of the earth was covered with black ashes, and we scarcely recognized it as the enchanting place of our former visit. At first we feared that this devastation had been caused by our own neglect; but on reaching the spot where we had encamped, which was separated from the surrounding hills by the rocky bed of the stream, we found the dry grass still around the place, which alone had escaped the fire. . . . A portion of the [Mexican] brigade [of General Carrasco] had passed the cañon a few days after us; and their twenty or thirty camp fires had, no doubt, communicated the flames to the grass, which had afterwards extended over the whole mountain (Bartlett 1854: II, 294–96).

[Clarke, May 27, 1849. On the Santa Cruz River at about the International Boundary.] After having traveled two or three miles, this morning, I discovered that I had left my belt and pistols at the camp. I rode back after them alone, as fast as I could urge my mule, although it was not very pleasant, as we saw signs of Indians near, and a corral they had used, and they are in the habit of lurking about camps. I found the grass about the camp on fire, and spreading rapidly (Clarke 1852: 83).

[Leach Wagon Road party, September 10, 1858, on the San Pedro River north of Benson.] In March 1858 the entire body of these lands were covered with a dense growth of sacaton grass averaging four feet in height and dry as tinder. Fire was communicated to it at a point about 20 miles below the site of Camp No. 14 and the entire length of the Valley of the San Pedro was traversed by the flames consuming every vestige of this once luxuriant growth. A much to be regretted attendant circumstance of this conflagration was the destruction of large quantities of Cottonwood, Ash and willow timber with which the banks of the river were densely overgrown. In three weeks after the occurrence of the fire it may be remarked that the Sacaton grass had grown up and covered the entire valley with a beautiful carpet of verdure (Itinerary 1858: 31–32).

The first and last of the six fires did not occur in the desert grassland, but in desert valleys, locations that so far as the writers know, no one has suggested were recurrently burned. In any event the evidence is overwhelming that mesquite and other shrubs were already well established in such locations. Two items, therefore, are irrelevant to the discussion.

Of the remaining four, three deal with fires started by the Caucasians themselves. The other records a distant fire; there is no way of knowing who started it, or whether it was in the mountains or in the grasslands. At any rate the emigrant train did not cross its wake, nor observe burned vegetation.

One is forced to conclude that these early journals do not mention a single mile on any of the traverses where the writers encountered grass that had been burned by a "natural" agent. They record three grassland fires, all of them set by the travelers. Such evidence as there is favors the reverse of the fire hypothesis; that the incidence of burning increased, not decreased, with the arrival of white men.

The latter hypothesis is not being seriously advanced, and there is, in fact, no adequate basis for a conclusion one way or the other.[10] One can state flatly, however, that the twenty-two accounts provide not the slightest historical justification for applying the fire hypothesis to the desert grasslands of this region, that they comprise the largest body of local historical materials yet examined in connection with the question, and that until some semblance of evidence is produced from primary historical sources, secondary authorities quoting the opinions of other secondary authorities will remain a poor substitute for fact.

THE AMERICAN CIVIL WAR AND AFTER

The outbreak of the Civil War brought to a close the American development of the Gadsden Purchase, such as it had been. Faced by what they thought to be imminent Confederate invasion, the Union garrisons at Forts Buchanan and Breckenridge burned their posts and fled to New Mexico, leaving the inhabitants to arrange for their own protection. The Santa Rita Mining Company closed its doors; the territory's first newspaper ceased publishing; Tucson, a center of secessionism, stayed within its walls while the Apaches ruled without, if indeed their reign had been interrupted at all (Eaton 1933, Pumpelly 1870).

The arrival of the California Column under General James H. Carleton restored Union authority (Keleher 1952), and in 1863 Lincoln signed the bill organizing Arizona as a territory. In December her first set of territorial officials arrived. Under them and Carleton normalcy returned. The years following Appomattox saw settlement again proceed. In 1867 homesteaders moved into the San Pedro Valley, bringing an end to its abandonment (Bell 1869: II, 92; *Report on Barracks* 1870: 464).

Through the 1870's as the army coped increasingly successfully with the Apaches, farming spread along the arable valleys; diggings multiplied in the mountains.[11] The year 1880 saw the arrival of the transcontinental railroad in Tucson. In 1881 a line connecting Benson, Arizona, with Guaymas, Sonora, by way of the San Pedro River, Babocomari Creek, Sonoita Creek, and Nogales was completed (Greever 1957).

The eighties witnessed the spectacular rise of Tombstone to her silver pinnacle as the Territory's first city—and her equally spectacular fall when water flooded the mines and the price of silver plummeted. The army found the final solution to the Apache problem; the "ancient scourge of Sonora" thereafter stayed on the reservation. By 1890, Arizona Territory had a population of 60,000 (U. S. Bur. Census 1895: XV). According to the assessors, 315,000 cattle roamed the ranges south of the Gila, and a figure twice that large is more realistic (Wagoner 1952: 121, *Report of Governor* 1896: 22).

The growth of stock-raising. By and large the story of the eighties is the story of ranching and its expansion out of Texas and Sonora onto the lush ranges of the desert grassland.[12]

Although there were only 5000 cattle in all of Arizona Territory in 1870 (U. S. Bur. Census 1872: III, 75), the following decade saw the foundations laid for a truly massive industry. Stimulated by government contracts for supplying beef to the military posts and the Indian reservations, large pioneering establishments like the Cienega, Vail, and Sierra Bonita ranches came into existence. Trail drives out of Texas and Sonora began, the excess above consumption going to stock the ranges. In 1872 Hooper and Hooker alone brought in four herds totaling 15,500 head (Wagoner 1952: 36).

In 1880 a Texas herd of 2500 grazed in Mule Pass east of the San Pedro; 3600 on the Babocomari Ranch; 3000 sheep along the San Pedro near Tombstone. The census figures show that the San Pedro Valley supported perhaps 8000 head of cattle and 10–12,000 head of sheep—more than the entire territory ten years before. South of the Gila there now were 20,000 cattle, and in the Territory about 35,000 (Wagoner 1952: 120, U. S. Bur. Census 1883: III, 141–42).

The 1880's witnessed a rapid expansion of the cattle industry throughout the West. The new railroads made it possible to raise beef for an eastern market; Indians had ceased to harass the ranchers; the depression of the 1870's had vanished. But the biggest factor, perhaps, was the ease with which the cattle industry attracted investment capital. A great ballyhoo campaign waged by railroad prospectuses, livestock journals, and territorial legislatures trumpeted to an eager public that the West held easy riches and that grass was gold.

A good sized steer when it is fit for the butcher market will bring from $45 to $60. The same animal at its birth was worth but $5.00. He has run on the plains and cropped the grass from the public domain for four or five years, and now, with scarcely any expense to his owner, is worth forty dollars more than when he started on his pilgrimage. . . . With an investment of but $5,000 in the start, in four years the stockraiser has made from $40,000 to $45,000. . . . That is all there is of the problem and that is why our cattlemen grow rich (Osgood 1929: 85–86).

So, instead of the West feeding the East, the East fed the West. In the scramble to stock the ranges and participate in the easy money, the normal flow of transport reversed itself and cattle from the feeder states along the Mississippi went back home. From 1882 to 1884 as many head traveled west as east (Osgood 1929: 92–93). As prices and profits soared, more eastern and English capital rushed to participate.

In Arizona by 1883–84 "every running stream and permanent spring were settled upon, ranch houses built, and adjacent ranges stocked" (*Report of Governor* 1896: 21). By the middle of the decade the governor could report that:

The number of cattle assessed this year has been 435,-000 and 50 per cent not assessed 217,500, making a total of 652,500 in the Territory at the present time (October 20, 1885). This number is being rapidly increased, and within another year it is expected that ranges with living springs and streams will be fully stocked (*Report of Governor* 1885: 8).

About the same time a reaction set in with the appearance of protective associations, formed by the established ranchers to protect the ranges against the flood of new cattle. They sought to adjudicate grazing rights by voluntary agreement and to restrict further immigration by means of quarantine laws aimed at keeping out "Texas fever" (*Southwestern Stockman,* January 10, 1885; March 14, 1885; April 11, 1885).

The organ of the Arizona ranchers, the *Southwestern Stockman* (February 7, 1885), wondered if the ranges were not overstocked, but concluded that the heavy recent increases were natural, and that the day of overproduction was "remote." The ranchers along the San Pedro Valley seemed less certain. In 1885 the Tres Alamos Association passed a resolution stating that the ranges were "already stocked to their full capacity," and demanding that the influx be controlled (*Southwestern Stockman,* July 25, 1885). A year later the Tombstone Stock Grower's Association spelled out the details:

Whereas, We, the members of this Association, perceive, with deep concern, that a crisis is fast approaching . . . demanding immediate action; and

Whereas, For several years past, there have been such large importations of cattle into Cochise county, from Mexico, and from our own States and Territories,

whereby these herds and their increase have so stocked our ranges, to the extreme limit of their capacity [that they] . . . leave us no surplus grass to tide us over . . . and,

Whereas, Sundry well disposed but misinformed persons, within our midst, have been, and are now, from time to time, circulating extravagant and highly colored reports concerning the resources of our county. . . . therefore be it . . .

Resolved, If, notwithstanding our remonstrance and protest, any person or persons persist in driving cattle into our section, without first legitimately securing sufficient grass and water for their herds, we will deny them all range courtesies (*Tombstone Daily Epitaph,* April 4, 1886).

But the time had passed when putting a stop to either courtesy or immigration would help. In anticipation of a better market the associations' own members allowed their herds to pile up (*Report of Governor* 1896: 22). The census of 1890 showed 1,095,000 range cattle for Arizona (U. S. Bur. of Census 1895: I, 29). The *Southwestern Stockman* (January 3, 1891) conceded at last that "the malady of overcrowding is with us in an aggravated form," and reported that disaster had been averted that summer only by the "phenomenal" late rains.

For 1891 the official assessment roll showed 720,000 cattle and the governor wrote that there were "closer to 1,500,000" (*Report of Governor* 1896: 22). The summer rains of 1891 were well below normal. In the arid foresummer of 1892 stock began to die. The summer rains of 1892 again were scanty, and by the late spring of 1893 the losses were "staggering" (*Report of Governor* 1896: 22). "Dead cattle lay everywhere. You could actually throw a rock from one carcass to another" (Land 1934). The governor estimated the mortality at fifty and possibly seventy-five per cent of the herds. At least for Pima and Cochise Counties the assessment rolls bear out his assertion (Wagoner 1952: 120–21).

That the overgrazing leading to the disaster of '93 had an enormous impact on the ecology of the region can hardly be doubted. Plate 43, among others, shows how the range looked. Thousands of square miles of grassland, denuded of their cover, lay bared to the elements. The cropping unquestionably weakened the old plant communities, leaving them open to invasion; it unquestionably upset the balance between infiltration and runoff, in favor of the latter. But whether overgrazing can be blamed solely or even principally for the events that followed is another matter.

The onset of arroyo cutting. Floods in the desert region can be documented well back into the Spanish period (Pfefferkorn 1949: 37–38, 41–42, 282; Thomas 1933; Wyllys 1931: 118), and it seems likely that they have been a part of the natural scene for as long as arid and semiarid conditions have prevailed. In some respects the climate necessitates their existence, for under the sparse, intermittent rainfall of the desert region a dense ground cover cannot develop. Without such a cover to impede runoff and promote infiltration, occasional rainfall intensities, particularly in summer, exceed the capacity of the ground to absorb water. Runoff results, reaching flood porportions in the event of unusually heavy or intense storms.

Although frequent floods used to occur (*Ariz. Quart.* 1880: 18; Hinton 1890: 112; Murphy 1928; *Arizona Weekly Star,* Sept. 12, 1878), and bridges became among the Americans' earliest construction projects (Conkling and Conkling 1947: II, 49; Ohnesorgen 1929; *Arizona Weekly Star,* Sept. 25, 1879), in general, before 1890, high water seems to have spread out in a shallow layer over the flood plains, causing some inconvenience but not much damage. Beginning about the mid-eighties the frequency and severity of the floods quickened.

In 1886 "the water in the San Pedro River was . . . higher than it was ever known to be. Between Contention and Benson there was four feet of water on the side of the [railroad] tracks" (*Arizona Daily Star,* August 30, 1886). At its junction with the Gila, "an avalanche of water swept down . . . like a wave, 6' high" (*Arizona Weekly Enterprise,* August 14, 1886).

On Sonoita Creek, in the same summer "water flowed down the valley some places 10 feet deep. At Calabasas the Santa Cruz overflowed its banks and swept a part of the valley" (*Arizona Daily Star,* August 3, 1886). In Nogales "the new adobe building of Mr. Samuel Brannon happened to be right in the course of a young river [and] . . . was totally destroyed, as was also his fish pond" (*Arizona Daily Star,* August 9, 1886). A Tucson newspaper reported that "some of the valley fields dived under water, and have not reappeared yet, they will probably be up in time to be assessed next year" (*Arizona Daily Star,* August 16, 1886).

During the following year, 1887, the San Pedro again had "higher water than . . . ever . . . known before" (*Arizona Weekly Enterprise,* Sept. 17, 1887). "For nearly the entire length of the river from Benson down to the Gila the crops with the exception of hay . . . [were] destroyed" (*Arizona Weekly Enterprise,* Sept. 3, 1887). At Charleston the flood carried away the dam shown in Plate 49 (*Arizona Weekly Enterprise,* July 16, 1887; *Arizona Daily Star,* July 17, 1887). At Nogales "the streets were a perfect sea of water and bridges and dams were no more than straws" (*Arizona Daily Star,* July 8, 1887). The Santa Cruz River at Tucson was "more than a mile wide and deep enough to float a mammoth steam boat" (*Arizona Daily Star,* July 15, 1887).

In 1888 and again in 1889, there were floods.[13] In 1890 events reached their climax, when within a two-week period at Tucson, the Santa Cruz carved a channel.

[August 5] The flood yesterday washed a deep cut across the hospital road, so that the road now is not only impassable but extremely dangerous for teams or travel as the embankment of the cut is perpendicular and the water below deep, and pedestrians might easily endanger their lives.[14]

[August 6] It is thought that the washout in the Santa Cruz, opposite this city, will reach Stevens avenue this morning. Boss Levine says that the Santa Cruz was higher last night than at any time during the last twenty-five years, and he ought to know as he has lived on its banks during that time.[15]

[August 7] The channel or cut being made by the overflow of the Santa Cruz river, is now one mile and a half long, by from one to two hundred yards wide—in other words—it extends from the smelter to about two hundred yards this side of Judge Satterwhite's place.[16]

[August 8] More than fifty acres of land which has formerly been under cultivation in the Santa Cruz bottom, has been rendered worthless by being washed out so as to form an arroyo.[17]

[August 9] The single channel which was being washed out through the fields of the Santa Cruz by the floods, resulted in considerable damage but this danger has been greatly increased from the fact that the wash or channel has forked at the head, and there are now several channels being cut by the flood, all of which run into the main channel. If the flood keeps up a few days longer there will be hundreds of acres of land lost to agriculture. As these new channels or washes are spreading out over the valley, they will cut through and greatly damage the irrigating canals.[18]

[August 13] The "raging" Santa Cruz continues to wash out a channel and the head of it is now opposite town. It may reach Silver Lake before the rainy season is over.[19]

On Rillito Creek, a tributary of the Santa Cruz, a flood during the same period "so cleaned out and deepened the channel that a third more water could be carried" (*Arizona Daily Star,* August 8, 1890).

On the San Pedro the chronology cannot be traced so precisely, but during the same two weeks, much the same thing seems to have occurred:

Of the country down the San Pedro, from Tres Alamos to the Gila [Captain Van Alstine] . . . says, "all of it is gone, destroyed, torn up, 'vamosed' down with high water." He never saw such a destruction in all his life. . . . The San Pedro never was as high as it was this time, and will not probably be for the next ten years. The losses sustained by the people will reach into the thousands (*Arizona Daily Star,* August 14, 1890).

At Dudleyville, near the mouth of the San Pedro the river "caved within 15' of Cook's [store]" (*Ari-*

zona Weekly Enterprise, Sept. 6, 1890), and up-stream at Mammoth flood washed the soil out in places thirty feet deep, exposing archaeological relics (*Arizona Daily Star,* Oct. 2, 1890).

The continuous present channels of these two rivers can be definitely dated from August, 1890. For the other streams of the desert region, the evidence is scantier and less reliable.

During the same month, a newspaper account states that San Simon Creek in eastern Arizona "was over a mile wide and running strong" (*Arizona Daily Star,* August 13, 1890). There is, however, no mention of channeling, and the commencement of the present steep-walled arroyo is a matter about which there is only secondary and contradictory evidence. It may have begun as early as 1883; it certainly had commenced by 1905.[20]

For the Sonoyta River [21] two hundred miles away on the Mexican Boundary, there is similarly no first-hand information. On the authority of an old inhabitant, Lumholtz states that there used to be a series of cienegas near the town, but that the flood of August 6, 1891, ripped a channel through them. "The swamps dried up in three years. . . . Where there had been before only a llano, a forest of mezquite trees sprang up" (Lumholtz 1912: 178–79). The channeling had certainly taken place by 1893 when the Boundary Commission noted that:

Sonoyta was formerly quite a flourishing little agricultural village, but heavy rains caused the river bed to sink so deep below the level of the surrounding lands that irrigation was attended with many difficulties, and . . . the village fell into decay, family after family moving away, until now scarce a half dozen Mexican families remain (U. S. Congress 1898: 23).

The relation between overgrazing and arroyo cutting. For many years it has been commonly asserted, sometimes on good grounds and sometimes not, that overgrazing initiated the cycle of arroyo cutting.[22] Whether it did is one of the knottiest problems with which the ecologist must deal.

The assertion that one caused the other is partly circumstantial and partly based on experimental observation. The circumstantial evidence stems from the close association of the two events in time. Large-scale grazing commenced in southwestern Arizona in the eighties. Flooding began in the mid-eighties and channel cutting, at least along the two major streams, in 1890.

The experimental evidence derives from observations by hydrologists that plant cover augments the amount of water infiltrating the ground, and that removal of the cover increases runoff and surface erosion. Supplementing this general relationship, which applies to the removal of cover by any agent—man, drought, or cattle—are field observations that indict man and his grazing animals more particularly. Well worn ruts—wagon roads, game and cattle trails, footpaths, drainage ditches—tend to become focal points for runoff, then rapidly enlarge because of their bareness and become arroyos (Duce 1918, Brady 1936). Thus, livestock have promoted erosion in two ways: by removing plant cover and by wearing trails. The explanation is plausible, but that does not establish it as fact; and two other considerations must be set against it.

The first has been argued by Kirk Bryan (1940) from geological evidence. Far from being unique, the arroyo cutting of the late nineteenth century was merely one of several erosion cycles that have visited the Southwest since late Pleistocene, alternating with periods of alluviation during which the channels filled and healed.

Arroyos similar to and even larger than the recent arroyos were cut in past time. As these ancient episodes of erosion antedate the introduction of grazing animals, they must be independent of that cause. . . . It seems reasonable to believe that the present arroyo is essentially climatic in origin. The introduction of grazing animals handled by optimistic owners may have reduced the already impoverished vegetation, and precipitated the event. Overgrazing thus becomes merely the trigger pull which timed the arroyo cutting in the thirty years following 1880.

Elsewhere (Bryan 1928a: 477) he reiterates his conviction that grazing was a secondary factor:

There is reason for believing that the fundamental cause of arroyo formation is a change to a drier climate. In this theory overgrazing becomes a mere accessory which set the date when cutting began.

Injecting geological considerations into the argument complicates it enormously, and has resulted in a multiplicity of views from which it is difficult to adduce any consensus. Over the years there has evolved a triangular range of opinion surrounding but somewhat off-center from Bryan's position, and with three apexes at complete cultural determinism, randomicity, and complete climatic determinism.

At one apex is the view held by Reagan (1924): that past periods of alluviation can be explained by aboriginal farming and dam building; that past cycles of channeling can be explained by the coming of great herds of herbivores—unidentified.

At another corner is the view of Dellenbaugh (1912), who states that he has noted erosion "in places where there were no cattle and never had been any," and who views channeling as continuous, not cyclic, occurring at irregular and independent intervals along individual streams as local circumstances dictate.

At the third corner is the view of Ellsworth Huntington (1914) who sees in arroyo cutting merely the inexorable shaping of land forms by the omnipotent hand of climate.

At various points within the triangle may be found a host of more sophisticated positions.

Antevs (1952) agrees with Bryan in dating the past cycles of cutting and attributing them to a drier climate. The present cycle, however, he attributes to overgrazing and asserts that:

Clearly, if left alone, the native vegetation could have weathered the droughts during which the arroyo erosion set in during the 1880's. The impoverishment of the plant cover which permitted the channeling must have been caused by new and foreign detrimental agencies, and the new factors during the 1870's and 1880's were large herds of cattle and sheep and numerous settlers. Therefore, the reduction of the vegetal cover which allowed arroyo-cutting to begin during the 1880's in the Southwest was caused by livestock and man.

Thornthwaite, Sharpe, and Dosch (1942: 127) tend like Dellenbaugh to regard past channeling as random and local, rather than systematic and regional in extent. Discontinuous channeling, they maintain, exists normally, and proceeds in slow waves up the course of a stream. An exceptionally big storm of high intensity may unite the segments, producing a continuous trench. If the vegetation is impoverished, "rains of . . . even moderate intensity can initiate a period of accelerated erosion."

By according vegetation this role, they admit overgrazing as a factor and emphasize it more strongly than Bryan. Whereas in his view "change of climate might be regarded as the debilitating agency that had lowered the resistance of the land—overgrazing as the germ or infection that had touched off the epidemic of accelerated erosion," in theirs: "overgrazing by livestock introduced by the white man has reduced the resistance of the land to erosion and . . . intense storm precipitation has been the germ or infection" (ibid., p. 123).

Hack (1939) and Luna Leopold have taken positions similar to Bryan's. Working at several places in the Rocky Mountain West, Leopold has refined the chronology for cut and fill, and has established almost beyond question its regional extent and cyclic nature. Recognizing a deficiency in Bryan's hypothesis—the failure to specify (beyond general aridity) the climatic factor that brought about the recent shift to cutting, he makes two suggestions. The first relates to the discontinuous trench segments that existed in 1880, and evidently had existed for some time:

The discontinuous channels and the lack of heavy alluviation during the period A.D. 1400–1860 possibly indicate an instability of the fill. The climate was not quite humid enough to cause further alluviation, nor was it sufficiently arid to cause degradation. On such a stage, postsettlement grazing could play a quick-acting and decisive role.[23]

He notes, secondly, that the weather records for New Mexico between 1850 and 1870 show a significant decrease in the number of small rains compared to the number of large rains, although there was no change in the average annual precipitation over the same period. This shift, he argues (Leopold 1951a), had an adverse effect on the plant cover.[24]

Although he clearly believes that climatic variation inaugurated the current cycle of arroyo cutting, Leopold recognizes that land use may either reinforce or counteract the effect of climate: therefore, "recognition of the current climatic variation does not require an abandonment of measures for improvement of land use" (Leopold and Miller 1954: 85). To this extent he occupies a position more adaptable than Bryan's.

Less optimistic, Judson (1952) maintains a viewpoint closer to Ellsworth Huntington's, although unlike the latter he associates arroyo cutting with periods of deficient rainfall: "An enlightened land policy will, of course, be important in preserving any gains made by nature. But it is extremely doubtful that even the strictest control of grazing, combined with 'upstream engineering,' will bring alluviation of the arroyos unless it is accompanied by sufficiently effective precipitation."

Gregory (1917: 132) remarks that: "For the Navajo country . . . human factors exert a strong influence but are not entirely responsible for the disastrous erosion of recent years. The region has not been deforested; the present cover of vegetation affects the runoff but slightly, and parts of the region not utilized for grazing present the same detailed topographic features as the areas annually overrun by Indian herds."

The same point is made by Peterson (1950: 421) who states that he has observed trenching in the Fort Bayard Military Reservation where grazing had been either excluded or rigidly controlled.

The particular climatic circumstances able to initiate channeling have also been a fertile source of controversy, even among those who hold climate responsible, and who recognize distinct periods during which cutting occurred. There is, first, disagreement as to whether secular trends or short-term variations are involved. Among those who favor trends, the consensus is that increasing aridity brought on the erosion, but the reverse has been widely asserted. Among the short-term variations suggested are periods of particularly intense and heavy precipitation; shifts in the relative amounts of summer and winter precipitation.[25]

The difficulty is twofold: first, in finding common ground for reconciling the interpretations produced by various disciplines working with dissimilar evidence; secondly, finding recent synoptic patterns that can be reconciled with the paleoclimatological record. The problem is not apt to be resolved soon to the satisfaction of everyone; nor is the related question of the role that overgrazing played in initiating the most recent cycle of arroyo cutting.

A second set of considerations arises from considering only the recent cycle of cutting, but extending the field of view to a larger geographical area. The two processes coincided in southeastern Arizona. Did they elsewhere?

The geographical extent of arroyo cutting. Kirk Bryan (1925) has reviewed the dates at which the current cycle of erosion began in Arizona, New Mexico, Utah, and Colorado and concludes that although channeling began at slightly different times in different parts of the area, the period from 1860 to 1900 encompasses nearly all of the cases.

Since Bryan's classic study, other evidence has come to light that tends to restrict the period still more. Cottam and Stewart (1940: 614) have demonstrated that erosion commenced at Mountain Meadow, Utah, during one protracted period in 1884. Gregory and Moore (1931:30 ff.) have fixed the period elsewhere in southern Utah in the eighties. Working further in New Mexico, Bryan (1927 and 1928b) has established dates in the late eighties for Rio Puerco and Rio Salado. In a study already discussed, Sauer and Brand (1931: 122–24) have concluded that in northern Sonora, a particularly critical area because of its early colonization, accelerated erosion began in the period from 1881 to 1891. Other workers have added to the assemblage.

Although dates for many localities still remain vague or unknown, and the last word has yet to be said, the twenty-year period from 1875 to 1895 evidently saw the beginnings of arroyo cutting from the Altar Valley, Sonora, to Kanab Creek, Utah; and from the Sonoyta River, Arizona, to Rio Puerco, New Mexico. The nicety of timing raises some interesting questions.

At Mountain Meadow channeling (1884) followed "settlement" (1862), and Cottam and Stewart (1940: 614) assume a causal relationship. On the basis of the dates given by Gregory and Moore, Bailey (1935: 350) predicates a similar relationship between settlement (1868) and channeling (1883) in the Kanab and Long Valleys of Utah. Colton (1937: 18–19) states that settlement (1878 or 1879) and overgrazing (the eighties) initiated floods (early nineties) on the Little Colorado.

But cattle-raising at Sonoyta dates from 1695—and accelerated erosion from 1891. "Settlement" in the Altar Valley of Sonora dates from the early 1700's, and erosion from about 1890. "Settlement" of the San Pedro Valley—depending upon which wave is meant, and assuming that one is talking only about Caucasians—can be traced to the 1760's when the Spanish presidio of Santa Cruz was founded, or the 1820's and 1830's when large-scale cattle-raising took place by Mexicans, or 1867 when

Anglo-American settlers came to the valley. Accelerated erosion dates from 1890.

Clearly one needs a clearer statement of what "settlement" is and what degree of "settlement" can be tolerated by a habitat without its becoming deranged. The historical evidence fails to indicate that the mere presence of Caucasians and cattle is enough to disrupt the equilibrium. Nor in light of the uneven pattern of conquest does it seem likely that the influx of settlers and their animals into an area of 300,000 sq. mi. could have been so nicely regulated as to result in a sort of permissive threshold of activity being everywhere crossed in one twenty-year period. Particularly when the two preceding centuries of colonization and grazing in the Mexican watersheds of the region did *not* see the threshold exceeded. The matter needs most careful study. The overgrazing that accompanied "settlement" may be the primary cause of arroyo cutting, or, as the initial historical evidence seems to indicate, it may be a mere auxiliary to broader factors.

The impact of cattle is easily perceived and appreciated: in the semiarid Southwest one can see it every day in the contrast between the grass on a fenced right-of-way and the adjacent range. But there is no similar easement through time along which one can observe the cumulative impact of small changes in climatic factors. That one can be readily perceived, but not the other, does not constitute a valid reason for assuming that only one is important.

Nor is the vagueness of the term "climatic change" a primary consideration. Although regrettable, it stems as much from our ignorance about the specific environmental requirements of plants as from an orientation toward determinism. In any event, its vagueness is matched by that of such concepts as "overgrazing" and "settlement." At the present time there is little reason for liking either set of terms; no justification for concluding that either cultural, or climatic factors exclusively inaugurated arroyo cutting; and little basis for ranking the two factors as primary and secondary. But if a choice must be made, the historical evidence, tentative though it may be, favors climate.

Culture, nature, and change. This chapter and the last one have presented the basic historical facts about man's occupancy of the desert region during the past three or four centuries, and have speculated about his ecological role during that time. The emphasis has been more on the activity of Indian, Spaniard, and Mexican than is customary, and rather less on that of the Anglo-American.

Although some imbalance is inherent in this presentation—certainly no one disputes the fact that Anglo-American culture has had a greater ecological impact than its predecessors—a treatment that emphasizes the others is overdue if for no reason except to counter the widely held view that before 1854 a state of nature prevailed in the Sonoran Desert, and the years since have seen the disruption of this natural order. There has been a continuum of cultural influence for as long as man has inhabited the region, and at times in the recent Spanish and Mexican past—well before 1880—disturbance by humans has attained a considerable magnitude.

In the first chapter, the principal climatic and vegetative features of the region were surveyed. Against the background of these two reviews—of "culture" and "nature"—the extent of vegetative change may be profitably examined. The next three chapters are devoted to paired photographs of the oak woodland, the desert grassland, and the desert, the three life zones in the changing mile above sea level. In preparation for a discussion of why the change has taken place, the plates establish some of the basic facts about its nature.

The Oak Woodland

IV

THE OAK WOODLAND

INTRODUCTION

The Upper Encinal. Within the life zone that he calls the Encinal, after the Spanish word for oak, Shreve (1915: 24–29) recognizes two subdivisions: the Upper Encinal occurs at elevations up to 6500 ft., or until it merges with the pine forest through an intermediate belt of vegetation called the pine-oak woodland by Marshall (1957).

Dominated by tree forms, mainly small oaks, the Upper Encinal has many affinities to chaparral vegetation: it is dense and largely closed; it contains a high proportion of shrubby species. At its upper edge the oaks are closely spaced and the intervals between them crowded with other woody members of the community: Arizona madrone, Wright's silktassel, Mexican pinyon pine, pointleaf manzanita. With decreasing elevation the underbrush thins; the oaks become more widely spaced; there is a tendency for grass to replace the shrubs; the Upper Encinal at length gives way to the Lower. The transition between the two marks the approximate upper limit of this study.

The Lower Encinal. The Lower Encinal, open and orchard-like in appearance, is found at lower elevations on gentler terrain. Here the large members of the community—oaks sometimes reaching heights of 80 to 100 ft. (Phillips 1912: 5)—are widely spaced, and interspersed with other trees like Arizona rosewood, one-seed juniper and mesquite. They are set in a matrix of grasses that include numerous gramas, as well as several species of three-awn, *Muhlenbergia,* and *Andropogon.* Several plants with succulent leaves or a succulent caudex occur with the oaks: beargrass, sotol, and species of agave and yucca. Small woody members of the community include skunkbush, mimosa, and fairyduster.

With lower elevations the Lower Encinal becomes more and more open, and the oaks are confined first, to relatively moist, cool sites on north-facing slopes and in canyons; finally, to canyon bottoms that at about 4000 ft. are separated from each other by ridges of grassland extending upward from the zone below. In this attenuated form the Lower Encinal ultimately merges completely with the grassland, or

in cases where that zone is missing, with the upper edge of the desert.

The gallery forest. Threading their way through Upper and Lower Encinal alike are thin lines of gallery forest or canyon forest, a distinctive vegetation that is restricted to watercourses and that retains its composition almost unchanged through the Encinal, across the grassland, and well into the desert. Goodding willow, velvet ash, cottonwood, walnut, Arizona sycamore, Texas mulberry, soapberry, and netleaf hackberry are its principal tree components. They may be accompanied by such shrubs as burrobrush, seep willow, and rabbitbrush, and by lianas like canyon grape and poison ivy. Closely confined to streamways, these plants span a greater range in elevation than most of the species on the adjacent uplands. The fact can probably be attributed to the relative homogeneity of temperature and moisture at all elevations, the habitats being streamways for cold air drainage as well as for surface and ground water.

The oak woodland. As used in this study, the term "oak woodland" or merely "woodland" means the same as "Lower Encinal," and will be used in preference to that designation. The oak woodland of the desert region is coextensive with similar vegetation eastward into New Mexico (Castetter 1956), and southward into the Sierra Madre of Sonora and Chihuahua (White 1948, Marshall 1957).

The plates that comprise the greater part of this chapter picture the oak woodland as it used to be and as it now occurs in an area of perhaps 2000 sq. mi., centered roughly around the Santa Rita Mountains of south-central Arizona. They follow the zone downward from an elevation of 5500 ft. to its attenuated remnants in isolated localities on north slopes at 3650 ft. The latter elevation is by no means the lower limit for the dominant woodland species. Both Emory oak and Mexican blue oak, for example, grow in Sabino Canyon in the Santa Catalina Mountains near Tucson at an elevation of 2850 ft. (Plates 70–72). There, however, they exist in the context of a gallery forest along a desert stream, not in the grassy setting that helps define a true woodland situation.

Four species of oak appear in the plates. The commonest is Emory oak (known also as bellota or blackjack), which may occur either in tree or shrub form, either in valley habitats or in the uplands. Mexican blue oak, a foothill species, is most common in the woodlands to the south, and can usually be distinguished at a glance from Emory oak by its sturdy whitish trunk, and the blue cast to its foliage. Arizona white oak, less common in the plates than the two species already mentioned, is also less xeric, and in the Lower Encinal is restricted to cool north slopes and valleys. Unlike Mexican blue oak, which never crosses the boundary between Upper and Lower Encinal, Arizona white oak ranges freely into the higher elevations. The last of the four species, Toumey oak, an unimportant species of limited distribution, typically grows as a shrub or a small tree, and appears here only in Plates 9 and 27.

The area that provides the setting for most of the plates is, in some respects, anomalous for its elevation. The annual rainfall is relatively high, with a high proportion falling as summer rain (65 to 70 per cent). The relatively mesic climate serves to pull down the elevational limits of the woodland, so that it occurs lower in this area than it does elsewhere in southern Arizona, although not necessarily lower than farther south, in Mexico (Marshall 1957: 32).

A comparison with Plates 42–60 shows desert grassland at comparable elevations along the San Pedro River. Rainfall seems to be the main deter-minant, but soil differences may in part be responsible; about the latter little is known except that limestone soils may appreciably elevate the zone boundaries relative to their location on soils derived from granite (Shreve 1942b: 192. Shreve 1922: 272).

Which plates to place in the category of oak woodland and which to assign to the desert grassland has sometimes been difficult to decide. Between any of the life zones there may exist an indeterminate expanse that is hard to classify. In the case of the grassland and the woodland the problem is complicated by other considerations as well: the pronounced interdigitation of the zones—upward along ridges and downward in valleys; the extensive displacement of boundaries under the ecological stresses of the past eighty years. An old plate may portray what is clearly woodland; in the corresponding new plate oaks may be completely absent. Toward the end of the chapter, grassland situations are extensively and increasingly pictured. On the other hand, oaks are mentioned in the chapters devoted to the other life zones.

In spite of these complications, the oak woodland presents fewer problems than either the desert grassland or the desert. It is easily recognized and relatively homogeneous wherever it occurs. As the pictures that follow will show, the secular trends within it are equally well defined.[1]

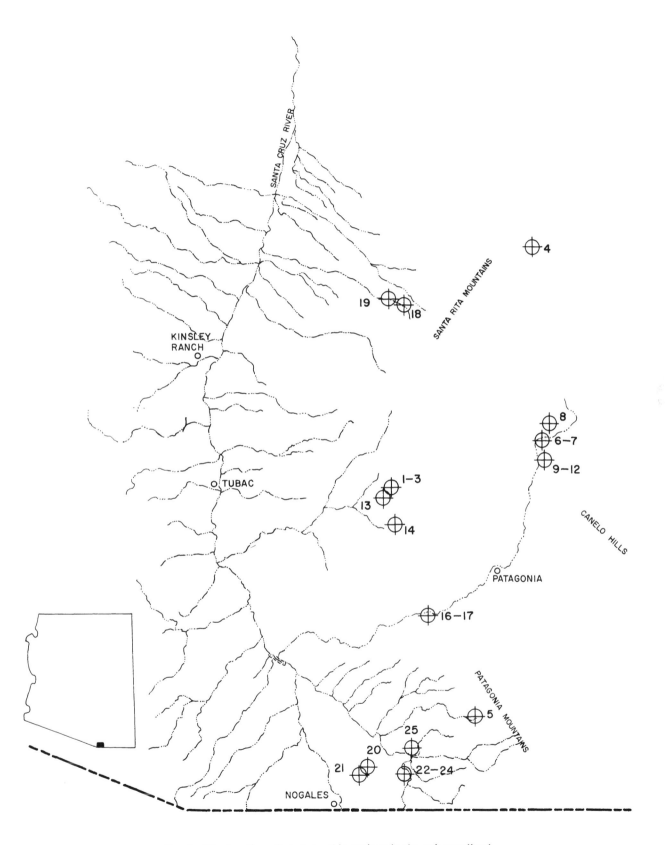

Fig. 7—The location of photographic stations in the oak woodland.

PLATE 1*a* (1891). The camera faces southwest toward a series of shafts and dumps at the El Plomo mine on Alto Ridge, on the southwestern side of the Santa Rita Mountains, also the site of the next two sets of plates. Local legend attributes the discovery of El Plomo to early Spanish missionaries, but similar stories have sprung up about most of the mines in this region, with equally little basis in fact. One can say with assurance only that the mine had been worked intermittently for at least fifteen years before this picture was taken (Schrader 1915: 197–98). The trees are oaks—either Emory or Mexican blue—and the relative scarcity of at least the former is probably due to woodcutting. Elevation 5500 ft.

PLATE 1*b* (1962). The ravine crossing the picture from right center to lower left runs northeast. The slope beyond it, therefore, has a north-facing component; that on this side, a south-facing. In the intervening years between the photographs, two mesquites have invaded the south-facing slope, but the opposing one is still clear. Two other mesquites can be seen on the skyline, but are actually located on the next south-facing slope over. At this elevation, the species is rapidly nearing the upper limit of its range (Parker and Martin 1952, Shreve 1915); one seldom finds it in colder, north-facing habitats. An alligator juniper occurs in the midground at the right near the ravine. The rest of the trees are oak.

PLATE 2a (1891). The camera looks north-northeast toward an oak-dotted saddle about one-fourth mile west of the diggings that appear in Plate 1. All of the El Plomo pictures, including six not reproduced, show oak to be more abundant today than in 1891; the size of the woodpile in the foreground suggests the reason. The stack is probably composed of Emory oak, preferred to the other woods of the region for fuel. Whereas Mexican blue oak and Arizona white oak were regarded as inferior, Emory oak fetched premium prices at the mines (Phillips 1912: 12–14, Swain 1893). The probability is slight either that mesquite, another good firewood, was abundant in the woodland, or that it would have been hauled in by burrotrain over the steep mountain trail leading to the mine. Elevation 5500 ft.

PLATE 2b (1962). The oaks today, although more numerous than in 1891, are restricted to the rocky slope at the right side of the picture. About three-fourths of them are Emory oak, and the remainder Mexican blue. The same proportion holds roughly for most of the south-facing slopes in the area although a third species, Arizona white oak, becomes important on the north slopes. The remaining trees in the picture are mesquite, and although none appears in the earlier view, here they easily outnumber the oaks. Throughout the woodland and the grassland, level saddles like that one in the midground have been breeding grounds for mesquite.

PLATE 3a (1891). A view looking north, showing a work crew walking along the trail that leads from the shaft in Plate 2 over the hill to the vicinity of Plate 1. All of the trees appear to be oaks. The scattered small clumps are probably beargrass, a likely source of the thatch on the lean-tos. Elevation 5500 ft.

PLATE 3*b* (1962). The trees are Emory oak, mesquite, and Mexican blue oak, none of which holds a clear dominance, and all of which are more abundant now than they used to be. The principal grass is sideoats grama. Among the smaller plants present are desert cotton, western coral bean, *mala mujer*, wait-a-minute, amole, and yucca. Two or three of the carcasses in the picture appear to be remains of trees in the earlier view.

PLATE 4a (1925). Along the old road between Tucson and So-noita looking southwest toward the north spur of the Santa Rita Mountains. The dominant tree in the photograph is Arizona white oak; a few Emory oak, one-seed juniper and mountain mahogany occur. In this locality the woodland is largely confined to ravines and north-facing slopes, a fact most evident, perhaps, in the background of the newer picture. Elevation 5100 ft.

PLATE 4*b* (1962). Although the two photographs span only thirty-seven years, they register some critical changes that may not be immediately apparent. Juniper has increased; oak has declined. Many carcasses can be seen in the picture, and of 134 oak trees, living and dead, between the camera and the dotted line, 40 per cent are dead. This figure indicates an extremely high mortality; unless the trend is reversed, the landscape in another few years will assume a quite different appearance. The highest death rate has occurred near the ridgeline: carcasses account for 60 per cent of the oak population on the upper one-third of the hillside; only 30 per cent on the lower two-thirds. In general, the invading junipers have sprung up beneath the canopy of existing oaks, a fact that may point to dispersal by birds. Although mesquite has invaded in the ravine at lower right, and on some of the sunnier slopes, the species is, as yet, of minor importance. Skunkbush and beargrass occur on the slopes; *yerba de pasmo* and burroweed on the ridges.

PLATE 5*a* (1887). In the foothills of the Patagonia Mountains east of the Santa Cruz River, about five miles north of the International Boundary, looking due east up Cañada de la Paloma. The tip of Guajalote Peak appears at the far right. A simple savanna vegetation prevails, made up of oaks and grass, with the trees confined to ravines and north-facing slopes. Emory oak dominates the bottomlands, but shares the upland habitats with Mexican blue oak. Elevation 4750 ft.

PLATE 5b (1962). A much more complex vegetation exists. Ocotillo has come in thickly on the south-facing slope at the left. Mesquite has invaded extensively along the ridges, and to a lesser extent on both the north- and south-facing slopes. In contrast to the situation at El Plomo mine, the winter temperatures at this lower elevation do not restrict mesquite to the warmer aspects. The number of oaks may be slightly less today than in 1887. Clumps of grass and *Mimosa dysocarpa* produce the tussocked appearance of the nearest hill.

PLATE 6a (About 1895). Looking north-northeast toward the north spur of the Santa Rita Mountains from a station three miles southwest of present-day Sonoita. In the midground Sonoita Creek winds through a trestle on the Fairbank-Guaymas railroad, completed in 1881 by Santa Fe and later sold to Southern Pacific. The trestle that appears here is about three years old, a flood in 1892 having destroyed its predecessor (*Arizona Weekly Star,* July 28, 1892). At the time of this photograph, erosion along the creek was still in its infancy. The relative length of the trestle timbers in the two pictures shows the extent to which channeling has progressed since. Apart from the grass cover, the vegetation on the hillside in the foreground is sparse. The trees are clearly oaks, probably Arizona white oaks. Although the shrubs cannot be positively identified, some of them at the lower right may be cliff rose. Elevation 4750 ft.

PLATE 6*b* (1962). At the time of this photograph the railroad track was being dismantled. The near hillside has a denser cover than before, including skunkbush, *1;* alligator juniper, *2;* Arizona white oak, *3; yerba de pasmo, 4;* beargrass, *5; Yucca (baccata* or *arizonica), 6;* a species of *Brickellia, 7; Rhus choriophylla, 8;* and mesquite, *9.* The dominant grass is sideoats grama. The areas at the right designated by letters have all been invaded by shrubs. In area *A* cliff rose predominates, with some mountain mahogany and *Ceanothus greggii.* Rabbitbrush occupies area *B* adjacent to the channel and, although the eroded flood plain of the creek is not visible, grows even more densely in the coarse alluvium there. Soapberry occupies the cliff above *B,* with Goodding willow in the stream bed nearby. In the steeper terrain in the rolling country beyond the track there has been less change. The oak population—Emory oak rather than the Arizona white oak of the hillside—is about the same as before. In the flatter areas and along the ravines, some mesquite has invaded. One large cottonwood, *10,* is visible in the left midground beside the railroad track. The hills in the background appear virtually unchanged.

PLATE 7*a* (About 1895). On Sonoita Creek looking upstream (southeast) from a point one-quarter of a mile below the trestle shown in Plate 6. Although the flood plain has already been dissected by erosion, neither mesquite nor seep willow has yet become established. The trees slightly right of center, growing near a small spring, are probably young cottonwoods, leafless for the winter. Channeling has occurred recently enough to disturb two of them. Elevation 4600 ft.

PLATE 7*b* (1965). Erosion has claimed the original camera station. Although this picture, looking through an opening in the mesquite bosque that covers the flood plain, is taken from a point perhaps thirty feet away from the old location, the same rocks on the hillside can be recognized. Seep willow fringes the creek on both sides, and is new on the scene. The midground vegetation at the right includes walnut, mesquite, netleaf hackberry, skunkbush, canyon grape, Goodding willow, Texas mulberry, Arizona white oak, and a group of cottonwoods, some of them probably the same individuals that appear in the old photograph. A large vine of poison ivy grows just off camera to the right. The vegetation of the north-facing slope in the right background is largely Arizona white oak, possibly with a sprinkling of Emory oak, an admixture of cliff rose, and along the crown of the hill, some mesquite, and a solitary one-seed juniper.

PLATE 8*a* (About 1895). Near Sonoita Creek, a mile east of the preceding station, and looking west-southwest across the ruins of Fort Crittenden, left and center midground, toward the south end of the Santa Rita Mountains. Crittenden, founded in 1867 near the site of Old Fort Buchanan, was closed in 1873, a generation before this picture was taken (Brandes 1960: 26–27). The corral belongs to a ranching operation and postdates the military complex. A pure, or nearly pure, stand of Emory oak meanders across the grassland following the course of a small runnel. Mesquites may be present, but if so, they cannot be identified. "For miles and miles the country is covered with grass and free of all sage, greasewood, or other worthless shrubs. Beautiful thrifty . . . oaks give much . . . beauty and charm" (*Arizona Citizen*, August 10, 1872). Elevation 4700 ft.

PLATE 8*b* (1962). The density of the adult Emory oaks has remained about constant, and in fact many individual trees are the same. However, the young oaks in the lower right-hand corner of the old photograph have not survived. Around the confines of the old corral, an appreciable mesquite invasion has taken place, a fact of some interest in connection with the cow-flap hypothesis. Downstream from the oaks, cottonwood trees, *1*, none of them evident in the old picture, dominate the scene, with a line of mesquites, also recent, at their base. New mesquites dot the pasture beyond the trees, and in the grassy foreground an incipient invasion is underway; arrows indicate the young mesquite plants.

PLATE 9*a* (About 1895). From a hill on the east side of Monkey Canyon about four miles southwest of present-day Sonoita, looking west across Monkey Lake toward the massif of the Santa Rita Mountains. Mt. Baldy and Mt. Hopkins are at left, Monkey Spring just off picture to the right, Sonoita Creek out of sight behind the midground ridge. The simplicity of the old vegetation contrasts dramatically with its complexity today. Oaks dominate the foreground; cottonwoods, leafless for the winter and still fairly young, grow around the lake. A few oaks can be seen on the valley floor near the spring. Elevation 4750 ft.

PLATE 9*b* (1962). Most of the oaks in the foreground have died and have been replaced by one-seed juniper, *1;* nevertheless, three oak species are still represented, Emory, *2;* Mexican blue, *3;* Toumey, *4.* Other shrubs are pointleaf manzanita, *5;* alligator juniper, *6.* The smaller plants include fairyduster, wait-a-minute, *Mimosa dysocarpa,* sideoats grama, slender grama, and a species of three-awn. In the rich valley flora are cottonwood, *7;* Arizona white oak, *8;* Goodding willow, *9;* red willow, *10;* velvet ash, *11;* Arizona sycamore, *12;* netleaf hackberry, *13;* coffeeberry, *14; Rhus choriophylla, 15;* desert broom; and a great many mesquite, all unlabeled. Along the road descending the ridge in midground: mesquite, juniper, and desert broom; elsewhere on the ridge: juniper, agave, and ocotillo.

PLATE 10*a* (1889). From the ridge in the midground of the preceding plate, looking west-northwest across Sonoita Creek and up Adobe Canyon toward the Santa Rita Mountains. Most of the trees are probably oak, although any positive identification is impossible. Note the open appearance of the Sonoita Valley. Elevation 4600 ft.

PLATE 10*b* (1962). The foreground illustrates a sight common throughout the lower woodland in this area: oak carcasses engulfed by living clumps of one-seed juniper. (Compare Plate 4*b*.) This picture, which shows only those foreground slopes with a south-facing component, may convey a false impression of juniper density. The species grows abundantly here on north-facing slopes, but seldom reaches the ridge lines or crosses them to the opposite face. The picture affords a bird's-eye view of a severe mesquite infestation. Below the dotted line—except for the cultivated fields—only a few cottonwoods and velvet ash in the valley, and oaks in the lower edge of the uplands have managed to compete with the tide of mesquite. The bosque is densest in the lowest part of the valley, progressively thinning toward higher ground. Above the dotted line, at about 4600 ft. elevation, mesquite becomes relatively unimportant. Oaks, velvet ash, and Arizona sycamore rule the canyon above the line; oaks, the uplands. Roughly the same demarcation can be detected in the old picture; there, however, it represents the transition between treeless areas and those with trees. Presumably the terrain above the line is relatively unchanged in its plant cover; that below, much changed.

PLATE 11a (1889). The picture looks south-southeast across
Monkey Spring (hidden from view behind the tree in left foreground) toward the north face of Mt. Hughes in the Canelo
Hills. The identity of the foreground trees is not clear, although
of the species still growing at the spring, they look most like
large *Rhus choriophylla*. The small clumps in the right fore- and
midground are probably beargrass. Below the spring the valley is
free from brush and appears to be a savanna dominated by oaks
and a few junipers. The foothills at the left look more open than
at present. Elevation 4600 ft.

PLATE 11*b* (1962). Mesquite has heavily infested the valley floor and the adjacent lower slopes. A dense thicket of cottonwood, Goodding willow, one-seed juniper, velvet ash, and *Baccharis neglecta* now lines the stream issuing from the spring (see Plate 12). In the foreground desert broom is the new dominant, although some beargrass still grows at the right (with dried inflorescences). In an effort to deepen Monkey Lake, a former owner bulldozed the bottom, inadvertently removing the seal that retained the water, and opening an intricate system of natural piping in the travertine underneath. The lake bed is now dusty and pitted, and the lower water table may account for a fair amount of the shrub and tree invasion.

PLATE 12a (1889). George Roskruge's surveying party standing above Monkey Spring. The camera site is perhaps twenty-five yards away from that of the preceding plate, and could be seen there if it were not for the trees. The view is slightly east of north, looking up the creek toward the spring, which is located at the foot of the hillock on which the group is standing. The stream banks, the valley floor, and the south-facing slopes beyond are almost devoid of large plants. The new picture stands in striking and, in many respects, puzzling contrast to this one. Elevation 4550 ft.

PLATE 12*b* (1962). The station is almost precisely the same as in the old picture, but the camera has been elevated to avoid a shrub. The barbed wire connects some fence posts that are in line with the old ones, but which are obscured here by shrubbery. The principal foreground plants are *Baccharis neglecta,* a relatively rare species; Goodding willow; one-seed juniper; *Rhus choriophylla;* Bermuda grass; watercress. The last-named two are exotics, not native to North America. In the midground: desert broom, canyon grape, *Baccharis neglecta.* Mesquite, Arizona white oak, and one-seed juniper hide the skyline, but in contrast to the valley floor and the stream banks, the surrounding hills look much the same. If, as seems likely, the spring has been continuously fenced off to avoid pollution, few of the changes immediately around it can be attributed to cattle.

PLATE 13a (1891). Ten miles northwest of Patagonia, Arizona, looking due east across Alto Gulch and up the western escarpment of the Santa Rita Mountains. The camera stations for Plates 1–3, 800 ft. higher in elevation, are situated about one-half mile away on the top of the ridge. The trees in the picture, some of which still survive, are Emory oak and Mexican blue oak. Almost no shrubs are present. Elevation 4650 ft.

PLATE 13*b* (1962). The large tree in the right foreground is a Mexican blue oak, as are most of the oaks on the opposite slope. A sprinkling of Emory oak occurs at the summit. Most of the trees, however, are mesquite. A careful inspection shows that all of the present-day oaks have counterparts in the old picture. No new ones have become established; mortality among the old has been severe. Hollow arrows mark four of the more prominent carcasses. The best survival has been registered in the rocky outcropping in the right mid-ground and at the higher, cooler elevations near the summit. The substantial thicket of ocotillo above the outcropping appears to be new. The small, dark shrub so abundant on the slopes is turpentine bush, also a recent invader.

PLATE 14a (1909). The old Salero mine in the southwestern foot-hills of the Santa Rita Mountains. The camera faces north-northeast; Salero Mountain is at the left. Many of the trees be-tween the camera and the buildings carry over into current times; all such relicts except the mesquites marked *l* are Emory oak. On the upper reaches of the mountains the vegetation is denser in the old view than now. Judging from the relicts, Mexi-can blue oak predominates there. Elevation 4500 ft.

PLATE 14b (1962). The buildings show less change than the plant cover. Of the trees in front of the dotted line, all are mesquite except the Emory oaks, 2; the Mexican blue oaks, 3; and one gray thorn, 4. Both species of oak have declined, the Emory, confined to the plain, and the Mexican blue, largely an upland type in this locality. In fact, no individual of either species can be seen that might not also have been present in the old photograph. Ocotillo, 5, has invaded the foothills. Although Mexican blue oak has decreased in density on the dry south- and southwest-facing slopes of the mountainside, mesquite has not appreciably increased.

PLATE 15a (1931). A classic development of the oak woodland about three miles south of Nogales, Sonora. The camera faces east-northeast. Emory oak predominates, although a few Mexi- can blue oaks may also be present. Arrows point to two young trees, an Emory oak and a mesquite, that are prominent in the later photograph. Elevation 4300 ft.

PLATE 15b (1962). In view of the relatively short period that has elapsed since the earlier photograph, the attrition among the oaks is remarkable. How much of the decrease may be due to wood-cutting, it is difficult to say. The mesquites (also a favorite fire-wood) have increased in number, and at this time of year, December, their partial defoliation makes them easily recognized. The dense grass cover in the earlier photograph has disappeared. Beyond the fence heavy grazing may be responsible, but on the semiprotected right-of-way in the fore- and midground, other factors must be operating. A partial explanation may lie in annual rainfall fluctuations. Note the advance of gullying on the other side of the railroad tracks.

PLATE 16a (1890). From a station seven miles southwest of Patagonia, Arizona, looking west toward what George Roskruge, the photographer, calls the Hill of San Cayetano. The Grosvenor Hills are at right; the San Cayetano Mountains, left. At this time the area evidently lay on the lower edge of the oak woodland. The trees are widely spaced and confined to ravines and north-facing slopes. A few junipers may be scattered among the oaks. The small, round tussocks are probably sotol. Elevation 4200 ft.

PLATE 16*b* (1962). Not a single living oak can be seen in the picture, although some relict Mexican blue oak can be found in sheltered spots nearby. Death has occurred recently enough so that an impressive number of carcasses still remain. None of them bears axe or fire marks, and as isolated as the area still is, any overt interference by man can be ruled out. Mesquite, the new dominant, shows much the same habitat preference as oak, the old; its greatest density occurs along ravines and on north-facing slopes. Another recent invader is ocotillo (right midground and scattered over the hill). Also present are desert broom, wait-a-minute, beargrass, Santa-Rita cactus, kidneywood, gray thorn, netleaf hackberry, a few one-seed junipers, and a species of yucca. The hummocks are composed mainly of fairy-duster and mimosa. Sotol has markedly declined, but is still common.

PLATE 17*a* (1890). Looking northeast toward Sanford Butte from the station used for the preceding plate. At the foot of the butte, Sonoita Creek is visible. If one makes due allowance for aspect, the vegetation is much the same as in Plate 16a: oaks along the ravines and on the one north-facing slope visible from this angle: sotol on the hillsides. In contrast to the conditions along its upper course (Plates 7 and 10), Sonoita Creek here has heavily vegetated banks, dominated probably by cottonwood. Elevation 4200 ft.

PLATE 17*b* (1962). All of the midground oaks are dead. Sotol still maintains itself on the rocky slope in the right foreground, but it has generally declined in importance, and in the foreground has been supplanted by beargrass. Although a few mesquites have come in, ocotillo is the most noticeable invader on the predomi-nantly south-facing slopes visible from this angle; it appears densest in those places where sotol has disappeared. The Sonoita Valley looks about the same. Cottonwoods still line the creek; the mesquite-studded spur indicated by arrows may be a little less open than before.

PLATE 18a (1911). On the northwest side of the Santa Rita Mountains, ten miles southeast of Continental, looking south-southeast up a woodland slope in the Santa Rita Experimental Range. Florida Canyon is at far left. Along the ravine in the foreground: oaks with an understory of young mesquites, seasonally defoliated. Young mesquites have also gained a hold on the lower part of the slope in the center foreground. Oaks dominate the hillsides; the compact, dark shrubs are sotol, probably interspersed with a few clumps of beargrass. In the distance at the far left the dense pine forest of the upper mountain region can be glimpsed. Elevation 4100 ft.

PLATE 18*b* (1962). The young mesquites in the older view now are adults and the fact that they have their summer foliage contributes to the cluttered appearance of the picture. A substantial population of ocotillo is present, all of it evidently recent. The numerous white stalks in the midground are the dried inflorescences of sotol. On the upper half of the hill at right, sotol and beargrass show a marked increase. The Mexican blue oaks along the wash in the foreground look much the same; intermingled with them are desert hackberry, Arizona sycamore, and low, spiny *Mimosa dysocarpa*. On the upper part of the hillsides the oak population remains unchanged, but downward the mortality increases, and many carcasses litter the mesquite woodland at the bottom of the slope in center midground. A number of mesquites have climbed the hills and some are, in fact, to be found almost at the summit at the right.

PLATE 19*a* (1911). A view looking southeast toward the ridge in Plate 18 from a camera station about a mile west-northwest of the preceding location, and at an elevation two hundred feet lower. Between the two stations the transition occurs between oak woodland and grassland. In fact, this photograph shows three life zones. From bottom to top: (1) grassland, already invaded by adult mesquites on the flats, but still open on the slopes; (2) the Encinal, ranging from the open oak woodland below, reaching down into the grassland on north-facing slopes and along ravines, to the dense Upper Encinal above; (3) the pine forest. Elevation 3950 ft.

PLATE 19*b* (1962). The old camera station cannot be recovered precisely, but the new one is within fifty yards of it. A dense mesquite thicket now dominates the foreground; the old grassland is nowhere to be seen. Other foreground plants, secondary but still important, are desert hackberry, catclaw, burroweed, and several species of prickly pear. Mesquite and ocotillo, both recent, dominate the low, formerly grassy hill at left center, and have engulfed a few blue paloverdes that carry over from the old view. Behind the hill and to the left is the only grassland slope that still remains pristine. Even there, mesquite, sotol, and beargrass have made some inroads. A tide of mesquite has inundated the lower valleys in Faber Canyon and has filtered up through the lower edge of the oak woodland.

PLATE 20*a* (1887). From a saddle on Proto Ridge, just east of Nogales, looking northwest across Proto Canyon toward Mt. Benedict, left center, and the Cerro Colorado Mountains, far left. Young mesquites are clearly recognizable around the monument. Oak dominates the runnels in the midground, but the intervening slopes are open and probably grassy. In the new picture *Yucca arizonica* is abundant, and many of the shrubby-looking plants in this view probably belong to that species. Elevation 4000 ft.

PLATE 20*b* (1962). From about thirty feet west of where the 1887 photograph was taken. The remains of the monument still exist, and the old station can be easily recovered; however, a dense stand of mesquite surrounds it, blocking the view. Fewer oaks (Emory and Mexican blue) grow along the drainage channels on the other side of the Nogales-Patagonia highway. Mesquite has replaced them, and dots many upland sites as well. Ocotillo dominates the south-facing hill at far right, and also grows densely on the steep southern exposures approaching Mt. Benedict.

PLATE 21*a* (1887). The view is similar to that in Plate 20, but the camera station is about one-half mile farther west. As in the preceding plate, oak dominates the vegetation, but its occurrence is restricted to relatively cool, moist sites in ravines and on favorably oriented slopes. The smaller shrubs are probably yucca. Elevation 4000 ft.

PLATE 21*b* (1962). The oaks (Emory and Mexican blue) are fewer, although one hardy survivor (an Emory) still stands beside the highway at left center. Mesquite dominates the ravines, and shares the slopes with *Yucca arizonica.*

PLATE 22a (1890). This is the first of three pictures taken by George Roskruge from or near the same hill on the María Santíssima del Carmen land grant just north of the Mexican Border in the Santa Cruz Valley. In 1890 the hill lay between the oak woodland and the desert grassland. In this picture, which looks southwest, aspect spelled the difference between the two zones. The oaks are sparse and restricted to the moister north slope. The south slope is pure grassland, devoid of shrubs and trees. The carcass of the mesquite marked by the hollow arrow still remains. The other two trees on the summit as well as the group of three designated by the solid arrow, are probably mesquite also. The remainder appear to be oak. Elevation 3950 ft.

PLATE 22*b* (1962). The lower edge of the oak woodland has shifted upward, leaving the area entirely within what remains of the grassland. No oaks at all are present. All of the trees in the picture are mesquite, which, besides dominating the north slopes, has markedly increased on the summit, and has invaded the southern exposure as well. The principal grasses are tanglehead and sideoats grama. Also present are wild buckwheat, fairy-duster, and a species of *Franseria*. In general the old division of oak on the north slopes and grass on the south has given way to mesquite on the north, light mesquite, and ocotillo (none of which is visible here) on the south.

PLATE 23a (1890). From a station part way up the hill from which the preceding plate was made. The camera looks north-northwest down the Santa Cruz Valley toward the San Cayetano Mountains at center. The Santa Cruz River enters the picture at lower right, bends sharply out of sight, and re-enters at right center, to follow a course marked by a line of cottonwoods. The tree at lower left is probably an oak, as is the one midway along the bare slope beyond it and slightly to the right. The terraces across the valley look like almost pure grassland. Elevation 3900 ft.

PLATE 23*b* (1962). The Yerba Buena Ranch now occupies the valley bottom, which has been extensively disturbed by cultivation. In the foreground: mesquite, prickly pear, grama grass, tobosa, fairyduster, three-awn, and *Amoreuxia palmatifida*. On the hills at left, mesquite with some sotol. Cottonwoods still dominate the stream banks; intermingled with them are yewleaf willow and Goodding willow. The arrow marks a relatively undisturbed section within the bend of the river where mesquite, catclaw, and gray thorn predominate. The terraces on the other side of the valley appear much brushier than before. The river channel is broader today and the bend appearing in the right midground sweeps farther left than it did in 1890.

PLATE 24a (1890). From midway up the east terrace of the Santa Cruz River looking west toward the hill from which the two preceding sets of photographs were taken. In the foreground in front of the field enclosed by the brush fence, the channel of the Santa Cruz—such as it used to be—can be seen. The trees along the stream appear to be mainly cottonwoods; oaks and probably mesquites occur on the other side of the valley. On the hillside the oaks are confined to the moister microhabitats. The trees on the skyline at the right of the picture appear in the lower left-hand corner of Plate 23a. Elevation 3800 ft.

PLATE 24*b* (1962). The oaks have disappeared; mesquite has invaded the slopes across the river, and in front of the camera. Although the Santa Cruz has not entrenched itself appreciably in the flood plain along this part of its course, its bed is wider and occupies the valley from the foot of the east terrace to the far side of the field pictured in the old view.

PLATE 25*a* (1887). The Santa Cruz Valley east of Nogales looking northeast from the river's first terrace toward Guajalote Peak, right, in the Patagonia Mountains. From individuals still surviving, the trees can be identified with reasonable accuracy. The wispier ones along the wash, center and right, are adult mesquites, and a single mesquite on the ridge is present in both photographs. The darker trees on the slopes and in the bottoms are one-seed juniper with a sprinkling of oak. Palmilla is also visible. Elevation 3800 ft.

PLATE 25b (1962). Mesquite has increased notably in the uplands as well as in the washes. An arrow marks the only remaining oak—an Emory. Juniper has decreased in density, but still thrives. Also visible are palmilla and gray thorn.

PLATE 26a (1931). Leaving Nogales, Sonora, which is located in the oak woodland, the Pan-American highway drops out of the uplands into the valley of the Río de los Alisos, and descends along that stream southward, encountering true desert vegetation near Imuris. In the relatively cool valley habitat, oaks extend downward beyond their normal elevational occurrence. Emory oak predominates near Nogales (Plate 15); at lower elevations along the highway, Mexican blue oak becomes in-creasingly important, until in the lowest reaches of the wood-land, it is the only oak species represented, and is confined to small patches on north-facing slopes. This photograph, taken about sixteen miles south of Nogales, shows one of the lower occurrences of the woodland in 1931. The grasslands in the valley already support a substantial mesquite population. Elevation 3650 ft.

PLATE 26*b* (1962). The oaks have disappeared; mesquites have migrated upslope into the old savanna; even where cultivation has not destroyed it, the grassland has deteriorated. The proximity of a small brick factory raises the question of woodcutting. Although some of the oak stumps bear axe marks, this does not necessarily mean that they were cut while alive. In the first place, oaks are difficult to kill; one would expect to find sprouts on some of the stumps if they had, in fact, been chopped down while living. Furthermore, Mexican blue oak, unlike Emory oak, makes an inferior firewood (Phillips 1912: 12). It is difficult to imagine a woodcutter climbing the slopes to cut them alive when living mesquites, equally easy to cure and an excellent fuel, were abundant on the valley floor. A likelier explanation seems to be that the oaks died and then, being already dry, were chopped down.

CONCLUSIONS

Changes in the oak woodland. The twenty-six sets of plates devoted to the oak woodland span a period of seventy-five years and reveal three general trends:

1) At all of the stations below about 4500 ft., and at a few above that elevation, the oaks have died faster than they have become established.

2) The mortality has been most severe at the lower edge of the woodland and, coupled with the failure of the oaks to repopulate at these lower elevations, has resulted in the migration upward of the boundary separating the woodland from the grassland.

3) Because of shrub invasion the woodland has become less open at all elevations.

The evidence suggests, without confirming, two other observations:

4) The mesquite invasion evidently antedates the decline of the oaks, which may be of fairly recent origin—dating from the 1930's or later.

5) The current decline of the oaks may be the most severe fluctuation they have undergone in several thousand years.

The mortality among the oaks. Plates 4, 9, 10, and 13–26 (17 out of a total of 26) show a severe attrition among the oak population. A census in Barrel Canyon (Plate 4) at an elevation of 5100 ft. reveals that 40 per cent of the oaks past the seedling stage are dead. Phillips (1912: 7) indicates an age upward of 150 years for Emory oak. Presumably Arizona white oak, the dominant species in Plate 4, has a similar lifetime; normal mortality, then, should at most be 1 or 2 per cent per year.

No counts have been made elsewhere that might serve for comparison, and the mortality at Barrel Canyon may not be typical of conditions at that relatively high elevation. Arizona white oak is less drought-resistant than Emory or Mexican blue oak. Down to about 4500 ft., there is no photographic indication that these latter, hardier species have declined. Below about 4500 ft. all of the plates, regardless of which oak species are present, show a decimation of the woodland. At 5500 ft., the uppermost elevation visited, both Emory oak and Mexican blue oak have *increased*. Woodcutting was extensive near these particular sites, however, and the oak population probably was decimated at the time of the first photographs, 1891. The apparent increase during recent years probably represents nothing more than a recovery from the earlier inroads.

The upward retreat of the woodland. At the sites shown in Plates 16 (4200 ft.); 22 (3950 ft.); 23 (3900 ft.); 24 (3800 ft.); and 26 (3650 ft.), the oaks have disappeared completely. In Plate 25 (3800 ft.) one remains. Although these locations lay at the lower edge of the woodland at the time of the older photographs, no observer would classify them as belonging there today. The boundary has shifted upward, leaving the localities behind, and in some cases the nearest oaks now occur several miles away at an elevation several hundred feet higher.

The invasion of the woodland. At all elevations the modern woodland is cluttered by species which, if present at all at the time of the earlier photographs, were less abundant. From the lower edge of the woodland up to 5500 ft. mesquite is denser. At least sixteen other species have invaded to some degree, at one or more of the stations.

As used here, the term "invader" is employed loosely, and includes three distinct categories of plants: (1) exotic species that have been recently introduced by man; (2) species primarily associated with lower life zones, but whose principal loci, like the lower boundary of the woodland, appear to be migrating upward; (3) species that were woodland plants at the time of the older photographs, but whose density has appreciably increased. A list of the plants, in addition to mesquite, that belong to these three categories is as follows:

Baccharis neglecta (Plates 11, 12)
Beargrass (17, 18, 19)
Bermuda grass (12)
Cliff rose (6)
Cottonwood (7, 8, 9, 10, 11)
Desert broom (11, 12)
Goodding willow (9, 11, 12)
Ocotillo (5, 13, 14, 16, 17, 18, 19, 20)
One-seed juniper (4, 9, 10, 11, 12)
Rabbitbrush (6)
Rhus choriophylla (12)
Seep willow (7)
Skunkbush (6)
Sotol (18, 19)
Turpentine bush (13)
Watercress (12)

Invasions by beargrass, cottonwood, Goodding willow, mesquite, ocotillo, and one-seed juniper have been observed at three or more stations, and it is safe to conclude that their increase has been fairly general. About the other dozen species, more information is needed. It should be noted that juniper is now less important than formerly in Plate 25, and sotol in Plate 17.

The date of the decline. The date at which the decline of the oaks set in cannot be fixed from the evidence available. The "before" photographs in Plates 4, 15, and 26 are comparatively recent, dating respectively from 1925, 1931, and 1931. Yet these plates show the decadence as clearly as the older pairs. The virtual absence in the literature of any references to oak decline also suggests that it is a recent event. Tree-ring studies now in progress may provide a definite answer.

The severity of recent trends. Plate 49 in the chapter dealing with the desert grassland, shows a particularly interesting situation. In the old picture, made about 1883, there appears an isolated patch of oak woodland in the middle of the San Pedro Valley above Charleston. Composed of Emory oak, it evidently was a relict dating back to the time when woodland bridged the valley from the Dragoon Mountains to the Whetstones.

The stand today, except for a single living tree, is dead. So far as is known the nearest living individuals of the species occur in the foothills of the Whetstone Mountains, perhaps ten miles to the west and almost 1000 ft. higher. There they are also associated with numerous dead and dying individuals; the lower boundary of the woodland in the Whetstones also appears to be retreating upslope.

If one could be sure that the Charleston patch of 1883 was, in fact, a relict colony, one could use its attrition during recent years to infer something about the magnitude of recent stresses. The fact that, having survived the vicissitudes of several thousand years of isolation,[2] the colony now is on the verge of extinction, would argue that recent events have been severe, indeed.

However, the possibility of accidental dispersal exists. The patch may not have been a relict, but merely a reintroduction by birds or animals. A study of tree rings may be useful here, and may illuminate the question of how severe recent stresses have been relative to others during the life of the oaks. The possibility remains that we are witnessing a change of no small magnitude.

The Desert Grassland

V

THE DESERT GRASSLAND

INTRODUCTION

Interpretations of the desert grassland, the next life zone below the oak woodland, vary widely from writer to writer, and the names used to designate it—if, indeed, it should be considered an entity and not merely a transitional stage away from the desert—are equally diverse.

The disagreement stems from three unresolved questions: first, the affinities of the zone to the short-grass prairie of the High Plains; secondly, whether the term "desert grassland" is not better reserved for the tobosa swales lying wholly within the desert; thirdly, whether the grassland is a true climax vegetation—that is, a stable association in equilibrium with the climate and soil of the region where it occurs—or whether it is an unstable, fire-induced form, susceptible to change even though soil and climate remain constant. No attempt will be made to review past interpretations, or to reconcile them with the discussion that follows.[1] As used in this study the term "desert grassland" applies to areas dominated either in the historic past or currently by grass, lying at a lower elevation than the woodland, but above or surrounded by the desert.

The elevational range of the desert grassland typically falls between 3000 and 4000 ft. It may go higher or lower depending upon microclimate, and it may be completely absent in cases where mountain masses rise abruptly from the desert floor. Thus, on the south side of the Santa Catalinas the upper edge of the desert merges directly with the oak woodland (Shreve 1915: 15), and there is no intermediate zone. Where bajadas sweep upward to the appropriate elevation, or where broad areas of gentle relief prevail, the grassland shows its best development. Such situations exist around many of the mountain ranges of the region, and at times of the year when the grasses are inactive the ranges are ringed by a conspicuous yellow sward that contrasts sharply with the belts of vegetation above and below it.

The grass flora of the zone is exceedingly rich, particularly at higher elevations. White (1948) gives a partial accounting of those found at the upper edge of the desert grassland of northeastern Sonora and lists 43 species; Shreve (1942b) names 48 species that occur in the grasslands of northern Mexico. Both lists could have been extended. Some variation has been noted with elevation, black grama dominating many lowland sites and blue grama, the higher elevations (Clements 1920, Darrow 1944, Wallmo 1955). The various species are not randomly distributed with respect to each other; nor are they uniformly spread across the expanse of the grassland. But what the distinctive mosaics may be, or where they occur are questions still to be answered.

Annual average precipitation ranges from about 12 to 16 in., with summer moisture slightly exceeding that of winter. Some of the grasses grow in March and April if rainfall and temperature are favorable. Most of the plants, however, are attuned to a cycle of summer activity, and their period of aestivation during the hot, dry spring is broken only with the onset of the summer monsoon (Nelson 1934).

At its upper edge the desert grassland resembles a woodland without oaks. At its lower edge it looks like an open desert—with, however, fewer shrubby species. In its maximum development it may occur either as uninterrupted shrub-free prairie—like the expanse around Sonoita, Arizona, today—or as a luxuriant matrix of grass studded by semiwoody plants like yucca, beargrass, and sotol. Woody species like fairyduster, Mexican crucillo, all thorn, and several of the mimosas and acacias dot the landscape in lesser developed areas. Their occurrence is apparently controlled by soil factors that are poorly understood.

As the plates will show, the desert grassland has undergone some significant changes in recent years. In the most altered locations grasses have been supplanted by a shrubby vegetation new to the area, at least in historic times. In this depauperate stage, the grasses are so scarce that they give hardly an indication of their past importance. Mesquite, acacias, burroweed, and other woody species dominate the landscape, with the intervening spaces barren of perennial plants.

For purposes of this discussion, the grassland has been divided into two geographical areas. The first fifteen plates deal with the belt centered around the Santa Rita Mountains. The final nineteen portray the grassland of the San Pedro Valley. Although the distinction between the two subdivisions is real, the species composition of the grass matrix is not necessarily different in the two cases. The eastern area has typically been invaded by Chihuahuan Desert plants, the western by Sonoran Desert species. The eastern has been invaded less by mesquite than by *Acacia* *vernicosa;* the western, not at all by the latter plant, and extensively by the former. Treating them separately makes it possible to perceive more readily the distinct development that each has undergone during the past three-quarters of a century, and on this historical basis the two subdivisions have been set apart.

The Grassland of the Santa Ritas

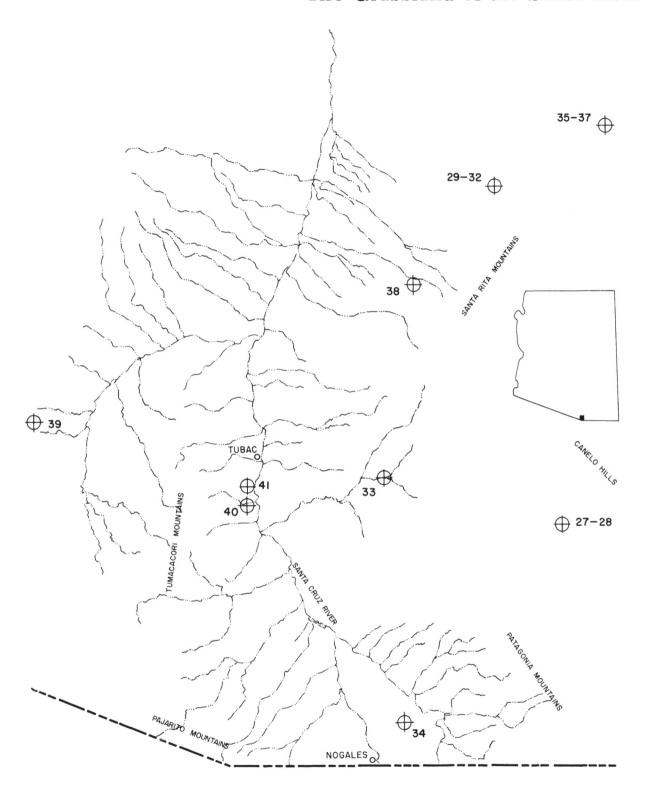

Fig. 8—The location of photographic stations in the grasslands around the Santa Rita Mountains.

[111]

PLATE 27a (1895). The camera, situated in the desert grassland, faces south-southeast across the valley of Red Rock Canyon east of Patagonia and records the vegetation mosaic of the region about 70 yrs. ago. Level areas and slopes with southerly aspect support grassland; the north slopes, oak woodland. In addition, on the hill at left midground, there is a dense chaparral vegetation which seems to be controlled not by slope and aspect, but by edaphic factors, since its occurrence is confined to island-like extrusions of shallow, red soil. In the valley the lowest terrace is probably covered by sacaton; the flood plain by burrobrush. Along the channel is found a scattering of deciduous trees, leafless for the winter, and probably including desert willow, mesquite, and velvet ash. A few large mesquites interrupt the grass cover of the upper terraces. Elevation 4450 ft.

PLATE 27*b* (1965). The most dramatic changes have occurred in the old patches of grassland, where mesquite now prevails. The patch of chaparral is apparently unchanged; it is dominated by Toumey oak and pointleaf manzanita and contains no mesquite. Much of the valley bottom is similarly free of mesquite; where the plant does occur here it falls within sharply delineated patches. On the slopes opposite the camera, the oak woodland, where most open, has been altered by the addition of many mesquites. The total number of oaks appears little changed, however.

PLATE 28*a* (1895). A view directly east from the station for the preceding plate. Here, because of the direction of the camera, it is possible to observe more clearly the sharp habitat preference of the oaks for ravines and north-facing slopes, and of the grassland for drier sites. What appear to be large mesquites grow along the base of the hill opposite the camera; large mesquite carcasses present on the hill in 1964 support the assumption that this plant was already well established there in 1895. The arrows mark two prominent plants that persist to the present; the open arrow denotes a Mexican blue oak and the solid arrow, an Arizona white oak. Elevation 4450 ft.

Plate 28*b* (1964). Aside from the notable increase in mesquite, two other changes, mostly out of camera's view, have taken place: of the many Mexican blue oaks standing in sheltered habitats along the ravine below the photo station, a high proportion are dead; secondly, small one-seed junipers grow beneath the oak skeletons, an invasion that is evident elsewhere in the panorama. (Arrows mark a few of the many junipers visible.) As in the preceding photo pair, the creek bed appears narrower and has more pronounced meanders in 1964 than in 1895.

PLATE 29*a* (About 1899). Looking west across the old mining camp of Helvetia, then in its heyday, from the northwestern foothills of the Santa Rita Mountains. Huerfano Butte stands alone at left center; across the Santa Cruz Valley, but not visible in the old picture, are the Sierrita Mountains. The ground cover in the foreground appears to be largely amole, the dried inflorescences of which are more prominent than the plants themselves. Interspersed with them are larger clumps of beargrass, a few yuccas, and, along the lower edge at the right, *Mortonia scabrella.* The dark trees at the left are Emory oak. A single ocotillo appears in the right midground. Note the open stretches of grassland and compare their appearance here with that in 1962. Elevation 4400 ft.

PLATE 29*b* (1960). Both mesquite and ocotillo have registered significant increases, the former along ravines and in the level grassland, the latter on south-facing hillsides and in the midground at the right. Most of the shrubs on the light-colored slope between the camera and the building with the pyramidal roof are *Mortonia scabrella*. This plant, too, seems to have undergone an increase, although as best one can judge, it was abundant in the old picture. The Emory oaks have grown, but otherwise are about the same. The foreground is still dominated by amole and beargrass, with sotol and *Rhus choriophylla* also present. Mexican crucillo, gray thorn, catclaw, and a few isolated blue paloverdes grow on the hillsides.

PLATE 30a (About 1899). Looking due west across the residential district of Helvetia toward the same hill that appears in Plate 29. The circles mark four shrubs that have carried over to the present day. The lower one is desert hackberry; the upper three, mesquites. This evidence that adult mesquite occurred in upland locations at this early date is of some interest. A few ocotillos can be picked out on the south-facing skyline. Mexican crucillo lines the wash at right. The ground cover in the foreground looks superficially like that in the preceding plate, but is actually different, consisting largely of yuccas, the inflorescences of which have been partly broken off or eaten. Beargrass evidently provided thatch for the houses, but no plants of the species can be seen in the picture. Elevation 4300 ft.

PLATE 30*b* (1960). The camera station is as close to the old one as the mesquite permits, but is nevertheless about twenty-five feet too far to the southwest. The north-facing slope in the foreground still has an abundance of yucca. Now, however, it is mixed with gray thorn, catclaw, and turpentine bush; and dwarfed by a second story of mesquite. The margins of the wash are, as before, dominated by Mexican crucillo, with some mesquite, desert hackberry, catclaw, and Wright lippia. On the wedge of higher ground lying behind the wash at the right: mesquite, Mexican crucillo, ocotillo, desert hackberry, catclaw, and snakeweed.

PLATE 31a (About 1899). Skyline retouched. Looking southeast across Helvetia toward the Santa Rita Mountains; the camera station for Plate 29 is in the left midground. Helvetia lies just below the border of the oak woodland. Oaks dominate the slopes of the Santa Ritas and, judging from relicts, reach down into the valley along the two prominent ravines. A few isolated individ- uals grow at favorable locations in the grassland. As was the case with El Plomo Mine (Plates 1–3), it is difficult to generalize about the tree population because of the presence of a settlement, in this case boasting a smelter (right), that must have burned large amounts of firewood. Elevation 4300 ft.

PLATE 31*b* (1960). The most pronounced change has been the obliteration of the grassland community by mesquite on the tablelands at the center and in the midground at the right. The shrubby growth along the wash in the foreground is also heavier, and includes Arizona white oak, mesquite, catclaw, gray thorn, desert hackberry, Mexican crucillo, *Sageretia wrightii,* and *Ceanothus greggii.* The hill in the midground at the left, a limestone intrusion, supports a community that is not typical of the area. Here, besides those described as foreground plants in Plate 29, Arizona rosewood, mortonia, and *Ceanothus greggii* are present. The Emory oaks scattered across the bajada seem to have changed little; many of the same individuals are present.

PLATE 32a (About 1899). Looking almost due north across Helvetia. The group of three tents at the right may also be seen in Plates 29 and 31. Of greatest interest at this early date is the presence in the fore- and midground of half-grown and adult mesquite trees. The slopes of the hill in back of town at the right of the picture appear open. Circles indicate eight of the larger shrubs that have carried over to current times. The uppermost is now dead. The one at the extreme right is a mesquite; the two at the left, Mexican crucillo; the others, probably Arizona rosewood. Elevation 4300 ft.

PLATE 32*b* (1962). The severe infestation in the lowlands needs no comment. Prickly pear, burroweed, and yucca are the most prominent plants associated with the mesquite. Amole forms the general matrix for plant life on the hill, and is responsible for the grainy, gray appearance of the slopes. Associated with it are several localized colonies of ocotillo, mortonia, and sotol. Ocotillo has probably increased although the lack of detail in the old photo makes comparison difficult. At the bottom of the hill a colony of mortonia occupies a triangular area that was much more open in the old picture, and that appears to have undergone more change than any other part of the slope. A single one-seed juniper, possibly present in the old picture, appears to the left of the triangle.

PLATE 33*a* (About 1891). A view of the Santa Rita Mountains looking east-southeast across a sloping valley dissected by many ravines. In the midground is Hacienda Santa Rita, the site of an early mining operation which was abandoned in 1861 because of warring Apaches; at the time of this photo, about 30 yrs. later, the camp has been reoccupied. Even at this early date mesquite is a conspicuous part of the upland vegetation. The carcass of the large mesquite (open arrow) is visible as a standing skeleton in the recent photo. Solid arrows mark what are presumed to be oaks although oaks are now missing from these spots. The tree with white branches in left foreground is a velvet ash which persists to the present. The branches in the right foreground arise from what is presumably a netleaf hackberry. Elevation 4150 ft.

PLATE 33*b* (1962). The collection of dwellings seen in the earlier view has all but disappeared among the many mesquites that now dominate the scene. What was formerly grassland sparsely studded with woody plants is now a woodland sparsely covered with grass.

PLATE 34a (1887). From Proto Ridge near Nogales, looking east across the Santa Cruz Valley toward Guajalote Peak, right center, in the Patagonia Mountains. The hills are conspicuously bare. On the nearest hill, which faces due south, a sprinkling of ocotillo can be discerned. Elevation 4100 ft.

PLATE 34*b* (1962). The camera station has been shifted twenty feet north in order to shoot through the ocotillo thicket now inhabiting the old site. The density of ocotillo has markedly increased on the south-facing hill in midground. Mesquite has invaded the ridges and, although not discernible in this view, has also invaded north-facing slopes and ravines; south-facing slopes are largely free of this plant. Also visible is the smaller fairy-duster. Sprucetop grama and sideoats grama are the principal grasses.

PLATE 35*a* (1915). From a hill west of Highway 83 between Tucson and Sonoita looking southeast across Davidson Canyon toward the highest peak in the Empire Mountains. In the foreground the dominant shrub is white thorn, with mesquite, ocotillo, an agave, a yucca, and two species of *Opuntia* also recognizable. In the midground, as far as Davidson Canyon, yucca and Mexican crucillo, identifiable from relicts, predominate on the uplands with some mesquite along ravines. Beyond Davidson Canyon the western bajada of the Empires is generally open, but, as relicts make clear, has a sprinkling of mesquite, white thorn, and yucca. Elevation 4100 ft.

PLATE 35*b* (1962). The fore- and midground vegetation is less open than before, but has the same constituents. In fact, many individual Mexican crucillo bushes are identical in the two pictures. Sotol, turpentine bush, and ocotillo can also be seen in the midground; tobosa, snakeweed, and sideoats grama occupy the spaces between shrubs. A few one-seed junipers can be found at isolated localities on the bajada; mesquite, catclaw, white thorn, and prickly pear are abundant there. From the mountain summit at 5500 ft. to its base at 4200 ft., a clear progression exists. The trees near the top are mostly Mexican blue oak, many of them dead, with an admixture of mesquite. The proportion of oak to mesquite decreases downward until the last oak, encircled, a dying Mexican blue oak, is reached about 4500 ft., and mesquite takes over completely. Clumps of beargrass dominate hillside *A*.

PLATE 36a (1915). Looking west-southwest toward Mt. Fagan, left center, in the Santa Rita Mountains, from a station near Davidson Canyon. Prickly pear and mesquite trees, many of them already adult, can be identified in the wash. The slopes ap-pear to be open desert grassland with a few mesquites and some ocotillo, but dominated primarily by Mexican crucillo, identifi-able from relicts. Elevation 4050 ft.

PLATE 36*b* (1962). Both the wash and the uplands are brushier. Along the drainage many mesquite trees are common to both pictures; an abundant and varied flora exists, including catclaw, white thorn, desert hackberry, netleaf hackberry, gray thorn, desert honeysuckle, burroweed, snakeweed, and Wright lippia. On the uplands Mexican crucillo enjoys about the same density as before, with a large number of individual plants carrying over from the earlier picture. The number of ocotillos has markedly increased, especially on rocky, south-facing slopes. Mesquite has also proliferated. The most important upland grasses are tobosa, sprucetop grama, and sideoats grama. Other common hillside plants are prickly pear, fairyduster, desert holly, and *Mimosa dysocarpa.*

PLATE 37a (1915). Looking south up Davidson Creek toward Mt. Baldy in the Santa Rita Mountains (right). Mesquites already inhabit the valley fill, although many of them appear still to be young. The hillsides at left and center in the background are moderately open, but with some shrubs, probably white thorn, mesquite, and ocotillo. Elevation 4050 ft.

PLATE 37*b* (1962). The construction of Highway 83 has badly disturbed the area. Of the relatively untouched sections, the hill in the background at the center is now heavily overgrown with ocotillo, white thorn, and mesquite. The bosque on the left side of the channel is made up predominantly of mesquite and catclaw, with a border of burrobrush. In the channel are seep willow, burrobrush, and desert broom. The shrub out of focus in the left foreground is white thorn. Russian thistle and a species of *Eriogonum* grow on the dike. In the right midground between the dike and the highway are mesquites, a walnut tree, burrobrush, snakeweed, and Russian thistle. In the two years that have elapsed since the site was first rephotographed, Russian thistle, a recent invader not native to North America, has become well established on the dike. At the turn of the century Griffiths (1910: 58) remarked that the plant was unknown in southern Arizona; now it is a common pest.

PLATE 38a (1911). From a ridge about one-half mile southwest of the camera station in Plate 19, looking southwest toward Elephant Head on the westernmost escarpment of the Santa Rita Mountains. The long outwash of the Santa Rita bajada drains left to right toward the Santa Cruz River, about eight miles away. It was in regard to this area at about the time of this photograph that Griffiths called attention to the mesquite invasion (see Introduction). In the foreground young mesquites and young ocotillos can be identified. Juvenile and half-grown mesquites have already become established in the grassland as well; by and large, however, the bajada still appears open and grassy. What may be oaks can be seen protruding from above the banks of the wash in the midground at the left (circles). Elevation 3900 ft.

PLATE 38*b* (1962). The new camera station is a few feet away from the old; however, a rock (Arrow 1) and the mesquite behind it can be recognized in both pictures. In the foreground the young ocotillo and mesquite have reached maturity. Kidneywood, catclaw, fairyduster, sotol, hopbush, *Mimosa dysocarpa, Opuntia* (*chlorotica?*), and several species of grama grass, and three-awn also grow on the hillside. On the bajada the proliferation of mesquite is obvious. The fairly open area marked by Arrow 2 now contains hopbush, desert broom, and sotol. Arrow 3 indicates a dense and recent invasion by ocotillo. The oaks, if that is what they were, have vanished.

PLATE 39a (1892). Looking north-northeast toward the Cerro Colorado Mine, worked by the Arizona Mining Company before the Civil War. "A prominent landmark for several miles . . . is the conical hill of reddish-colored rock called . . . the 'Cerro Colorado.' . . . Standing on a rise of rolling land, isolated from the neighboring mountains, it presents in its conformation and coloring a singularly picturesque feature in the scene. Back of this curious peak to the north lies a rugged range of mountains, upthrown, as it were, out of the earth by some tremendous vol- canic convulsion. . . . The headquarters lie on a rise of ground, about a mile distant from the foot of the Cerro Colorado, and present at the first view the appearance of a Mexican village built around the nucleus of a fort. . . . The works are well pro- tected by a tower in one corner of the square, commanding the plaza and various buildings and storehouses, as also the shafts of the mine which open along the ledge for a distance of several hundred yards" (Browne 1951: 265–66). Elevation 3650 ft.

PLATE 39*b* (1962). "At the time of our visit it was silent and desolate—a picture of utter abandonment. The adobe houses were fast falling into ruin; the engines were no longer at work" (Browne 1951: 266). The rock formation in the right foreground of the old picture appears to the left of its old location and farther away from the camera. Although the grass on the foreground slope is still dense, the grassland in general is shrubbier. Among the invaders are ocotillo, catclaw, and mesquite.

PLATE 40*a* (1891). Looking southeast across Tumacacori Mission from a station on the west terrace of the Santa Cruz River. The San Cayetano Mountains are in the background. "The mission . . . is pleasantly situated on a slope, within a few hundred yards of the Santa Cruz River. A luxuriant growth of cotton-wood, mesquit, and shrubbery of various kinds, fringes the bed of the river and forms a delightful shade. . . . Tumacacari [*sic*] is admirably situated for agricultural purposes. The remains of acequias show that the surrounding valley-lands must have been at one time in a high state of cultivation. Broken fences, ruined out-buildings, bake-houses, corrals, etc., afford ample evidence that the old Jesuits were not deficient in industry. The mission itself is in a tolerable state of preservation" (Browne 1951: 152). From the open condition of the terrace slopes, it is apparent that, away from the valley, the desert grassland prevailed, even at this low elevation of 3350 ft.

PLATE 40*b* (1962). The mission, now a National Monument, appears to be in a better state of preservation than formerly. Apart from recent cultivation, the valley floor looks about the same. The terrace slopes, however, are now dominated by white thorn and jumping cholla, and the open, grassy appearance is a thing of the past. Prior to the old picture Tumacacori had been a mission center for two centuries. Eighty years before the old picture it was the center of extensive cattle raising on the Tuma-cacori and Calabasas land grant. It is doubtful that in this vicinity either the human population, or the livestock population, or the degree of cultural disturbance in general during Anglo-American times has attained the level that it reached in Mexican and Spanish times. The extensive vegetative changes that have taken place on the terrace since 1892, then, must be attributed to factors other than cultural.

PLATE 41*a* (1890). This view, taken little more than a mile north of Plate 40, looks east across the Santa Cruz River toward (from right to left) the San Cayetano Mountains, the Grosvenor Hills, and the Santa Rita Mountains. In the foreground, to the left of the rock monument, is white thorn, a shrub of the uplands here and, by inference, across the valley as well. At the base of the hill in the foreground, mesquites and other shrubs among which are probably catclaw and gray thorn. The Santa Cruz River is lined by cottonwoods and probably mesquite. On the far side of the valley is a dark band of vegetation that coincides with a present-day mesquite bosque. The low hills across the valley appear dark as though covered by shrubs. Elevation 3250 ft.

PLATE 41b (1964). The most conspicuous changes have to do with buildings and cultivated land. Carmen, Arizona, and its cotton fields are new, but little change has taken place in the native plant cover during the past 74 yrs. The mesquite bosque across the valley is no more prominent today than before, and the many large trees present there today attest to its great age. Any changes that may have occurred on the hills across the valley, densely covered with white thorn today, are not easily discerned, but the small area of foreground hillside that is subject to comparison shows an increase in shrubs, mainly white thorn. No cottonwoods appear along the present entrenched channel.

The Grassland of the San Pedro Valley

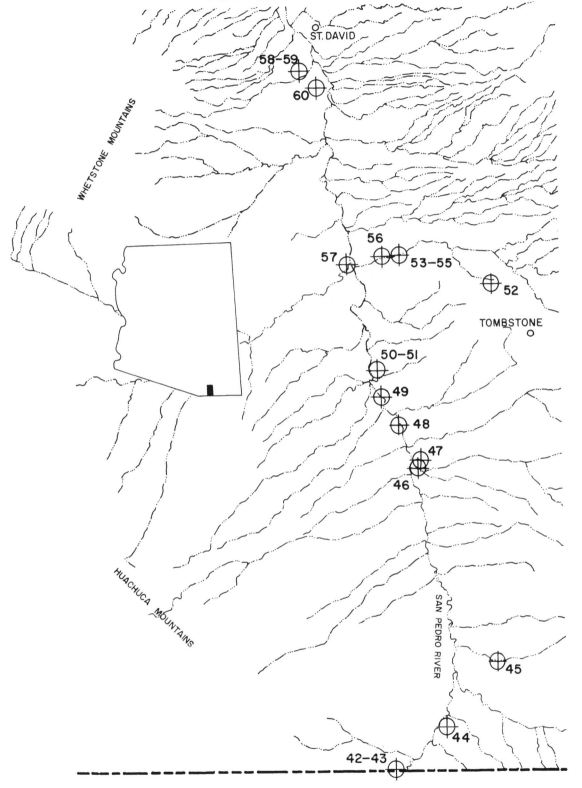

FIG. 9—The location of photographic stations in the San Pedro grasslands.

PLATE 42a (1891). On the International Boundary at a point just west of the San Pedro River, looking southeast into Mexico toward the Sierra San José. "At this point, approaching from the east, the traveller comes within a mile of the river before any indications of a stream are apparent. Its bed is marked by trees and bushes, but it is some sixty or one hundred feet below the prairie, and the descent is made by a succession of terraces. Though affording no very great quantity of water, this river is backed up into a series of large pools by beaver-dams, and is full of fishes. West of the river there are no steep banks or terraces, the prairie presenting a gentle ascent" (Emory 1857: I, 99 ff.). The slopes of the terrace along the east side appear, like the valley, to be grassy and open. Palmillas inhabit the basin in front of the camera, and several individuals among them persist into the present. The other shrubs may be Mexican tea or, in the case of the larger ones, mesquite. Elevation 4350 ft.

PLATE 42*b* (1962). *Acacia vernicosa* instead of grass dominates the terrace sides. The river's deeply scoured channel is marked by a line of trees that were not identified, but which, in addition to cottonwood, may include Goodding and desert willow, velvet ash, and walnut. A dense growth of mesquite and sacaton grass blankets the flood plain. In the small basin: palmillas, some of them probably a century old, mesquite, desert willow, rabbit-brush, and sacaton. The curved spit of high ground half encir-cling the basin supports Mexican tea, wait-a-minute, threadleaf groundsel, mesquite, fairyduster, blue grama, hairy grama, *Aris-tida glauca*, and a species of blue stem.

PLATE 43a (1891). From a point fifty feet east of the preceding station and looking west up the bajada of the Huachuca Mountains along the International Boundary. The wash at the left bends abruptly southward, and is at the right in Plate 42. In it palmillas can be recognized and, to the left of and behind the surveying party, what appears to be a grove of desert willows. The scarcity of grass is probably not representative of the conditions before 1890. The year 1891 marked the peak of overgrazing on Arizona ranges; in addition, the summer saw the beginnings of a disastrous drought that decimated cattle herds all over the region. Emory (1857: pt. 2, p. 18), thirty-seven years before this picture, described the bajada as "composed of hard gravelly soil, and supporting a close sward of grama grass, giving a peculiarly smooth shorn look to the general face of the country." Elevation 4350 ft.

PLATE 43*b* (1962). This picture was taken on the Mexican side of the fence and looks toward Monument 98, which has replaced the old rock cairn. The foreground, almost unchanged, supports wait-a-minute, several small mesquites, and some threadleaf groundsel, a plant that is seldom grazed and is toxic to cattle. Although mesquite, desert willow, and desert broom choke the wash, woody plants have invaded the uplands only slightly. The grasslands of the upper bajada here and those of the rolling country around Sonoita are the only major ones in southeastern Arizona that remain free from brush and substantially unchanged. In the background, at the foot of Montezuma Peak and extending down into the bajada along drainages, is the oak woodland.

PLATE 44*a* (1891). On the east side of the San Pedro River about two and one-half miles north of the Mexican Boundary looking southwest; at far right are the Huachuca Mountains. The low hills in the midground support a number of shrubs, some of which persist into modern times. Among them are one-seed juniper and *Rhus microphylla*. On the basis of their resemblance to the modern vegetation the abundant dark plants between the hills and the little valley in front of the camera can be tenta-tively identified as Mexican tea and all thorn. Mixed with them are a number of palmillas, two of which appear to carry over into the new picture. To the right of the hills and beyond the house is a light-colored, dense field of sacaton grass which has also persisted, largely unchanged, down to the present. In it are a few scattered trees, possibly cottonwoods, growing near the San Pedro River. Elevation 4250 ft.

PLATE 44*b* (1962). Two species account for most of the changes: mesquite has invaded the little valley in front of the camera; *Acacia vernicosa* has moved into the rolling country behind it and onto the hillsides, imparting to both areas a characteristically dark cast. On the foreground hill: curly mesquite, black grama, and sprucetop grama, all grasses. In the valley: mesquite, palmilla (with the conspicuous white inflorescences), rabbitbrush, tobosa grass, and sacaton. In the uplands between the valley and the hills: *Acacia vernicosa, Rhus microphylla,* and oneseed juniper. Two large cottonwoods can be seen at the far right, slightly behind the hill.

PLATE 45*a* (1891). The north end of the Huachuca Mountains as seen from across the upper San Pedro Valley. The camera looks west down Spring Creek across the broad valley which along this section supports shrubby vegetation alternating with more open areas. The photo station is located in one of the lat- ter; a sparse covering of grasses is interrupted by plants of blue yucca, palmilla, and sotol. Although positive identification is not possible, shrubs seen in the photo probably include creosote bush, Mexican crucillo, fairyduster, and *Acacia vernicosa*. Elevation 4400 ft.

PLATE 45*b* (1962). After seventy years the patches of grassland are gone and sites already shrubby before are even shrubbier now. The camera station is covered by *Acacia vernicosa;* the small valley below the camera supports a growth of mesquite, desert willow, *Rhus microphylla,* and sacaton. The three relatively large trees growing along the channel at center midground are walnuts.

PLATE 46a (1891). From a point on the east side of the San Pedro about one and one-half miles southeast of Lewis Springs, looking north toward the Lewis Hills. The picture shows a classic grassland community, palmilla being the large dominant. Because of the dead, inflammable material along its stem the plant is highly susceptible to fire injury, which, while not necessarily fatal, does induce a different form (inset), with multiple sprouts being sent up from the uninjured subterranean stem. Many of the yuccas in the main picture are over fifty years old, have shaggy stems, are not clumped, and in short do not show evidence of having been burned. It seems clear that recurrent fire cannot explain why mesquite and other woody invaders are absent from the area. Elevation 4050 ft.

PLATE 46*b* (1960). From a station within one hundred yards of the old one, which cannot be located precisely. The railroad cut and the dirt highway to Lewis Springs have disturbed the surface so badly that it is almost unrecognizable, but in spite of the disturbance, three major invasions can be distinguished: beyond the highway, *Acacia vernicosa* has replaced the grassland; a mesquite bosque dominates the lower-lying area between the acacia and the river; cottonwoods line the channel of the San Pedro. In the foreground at the right *Acacia vernicosa;* at the left, the dried inflorescence of a palmilla. In and along the cut: sacaton, Mexican tea, gray thorn, mesquite, and *Acacia vernicosa.*

PLATE 47a (1891). The picture was taken about one-half mile north of the preceding one and looks in the same direction. Here, however, the camera station is on the border between the yucca-grassland community of Plate 46 and the low-lying plain surrounding the San Pedro River, which is about one-half mile away at left and center. A palmilla is in the foreground, and three cottonwoods can be seen in the midground. Burrobrush and small mesquites may be two of the other midground plants. Elevation 4050 ft.

PLATE 47*b* (1962). Only mesquite branches can be seen from the old station. The new one is on the top of a dike within fifty feet of the old, and looks out across the top of the bosque that now blankets the plain. Among the treetops at left is a relict palmilla. The inset is a close-up view of the plant, which, in competing with the mesquite, has grown to an atypical height. On the floor of the bosque: sacaton, burroweed, desert holly, cane cholla, Mexican tea, and palmilla. Burrobrush occupies some areas closer to the river. The cottonwoods, partly leafless in December, have greatly proliferated since 1891.

PLATE 48*a*. (1891). The San Pedro River, visible between the second line of low trees and the railroad tracks, follows what appears to be a channel with already eroded banks. The course is difficult to follow, partly because of the absence of the large riparian species that mark it today. Although the valley and the hills beyond are relatively open, at this date they already support a prominent shrub stratum, varying greatly in density. Elevation 4050 ft.

PLATE 48*b*. (1962). The river channel, now outlined by cotton-wood, willow, and mesquite, is deeply entrenched in the valley just beyond the cultivated fields. The hills to the east are covered by shrubs among which are *Acacia vernicosa*, creosote bush, and ocotillo.

PLATE 49a. (About 1883). A few miles downstream from the preceding plate, looking east-southeast toward Bronco Hill and the Gird Dam, one and one-half miles above the old town of Charleston. The dam, which supplied water to the mills at Charleston for use in processing the ore from the Tombstone mines, was destroyed in 1887 by a forerunner of the flood that three years later initiated channeling on the San Pedro. The en- circled tree at the upper left is an Emory oak, one of the now extinct colony whose carcasses still litter the moister slope on the other side of the ridge.[2] Their presence in the valley, even in 1883, is surprising; they can be best explained as a relict colony left from the time when oak woodland bridged the San Pedro Valley from the Mule and Dragoon Mountains on the east to the Huachucas and Whetstones on the west. Elevation 4000 ft.

PLATE 49*b*. (1960). The modern steep-walled channel is obscured by trees; however, erosion has claimed nearly all of the ground on which the wing of the dam rested, at the left side of the picture. Cottonwood and Goodding willow now line the river; seep willow and Mexican devilweed grow abundantly on the sandy bed. The rocky hill in the foreground supports a distinctive community composed of Wright lippia and *Acacia vernicosa*, an association which also dominates the hills in the right midground and the left background. On the slopes of Bronco Hill, center, sotol joins the other two plants. In all locations the vegetation is appreciably denser than before. Along the floor of the little valley in front of the camera, mesquite and *Acacia vernicosa* have invaded. Also visible in the picture are ocotillo, desert broom, chamiso, soapberry, and sacaton grass. The oaks have nearly vanished; except for one survivor the nearest ones, so far as is known, occur perhaps ten miles away and nearly 1000 ft. higher in elevation.

PLATE 50a. (1884). From a hill north of Charleston looking south-southeast across the town toward Bronco Hill. The San Pedro River is off-picture to the right. At this time Charleston was in its heyday, boasting two mills (the Gird, left midground, and the Corbin, left background) and a polyglot population of some two thousand. Properly speaking the town shown here is Millville; Charleston, its commercial and residential sister, lay across the river. Gird Dam, in the preceding plate, supplied the mills with water. The flumes in this picture, in turn, conveyed tailings to the pond shown in the next plate. Ecologically, the point of most interest is the open, grassy appearance of the rolling country on the far side of town. Charleston at this time clearly lay in the grassland. Elevation 4050 ft.

PLATE 50*b*. (1960). Of man's work only the mill foundations and the adobe walls of the old Gird house remain. The foreground grass has given way to a thicket of *Acacia vernicosa,* with tanglehead and sideoats grama occupying the spaces between shrubs. The midground, center and right, is dominated by mesquite, with an understory of sacaton. The hill behind the Gird house, a rocky habitat similar to those in the preceding plate, supports the same association of Wright lippia and *Acacia vernicosa*. In the darker area at center and immediately to the right of the old house, the lippia drops out, leaving a pure stand of acacia. In the lighter, more open area to its right: small mesquites, sacaton, and *Aristida glauca*. On the rolling country in back of town the grassland has been engulfed by a dark tide of *Acacia vernicosa* studded locally with patches of tarbush and creosote bush. The light area (arrow) at extreme right is perhaps a relict colony of the old grassland; in it occur scattered mesquite, palmilla, Rothrock grama, and *Aristida glauca*. Also abundant in the picture are cane cholla and catclaw.

PLATE 51*a* (1883). Charleston, for eight years one of the principal cities of Arizona. The picture is taken from the hill that appears between the camera and the Gird Mill in the preceding plate, and the view is downslope, southwest toward the Huachuca Mountains. The San Pedro River runs from left to right through a channel trench that is well defined, but hardly comparable to the modern one. The foreground hillside is spotted with *Acacia vernicosa.* One lone cottonwood stands on the riverbank; the slopes of the terrace across the river are open and relatively free from brush. Elevation 4000 ft.

PLATE 51*b* (1960). The old city has vanished. Its death knell sounded when the Tombstone mines struck water, making the nine-mile ore-haul to the river unnecessary. A drop in silver prices accelerated the decline of both towns. Finally, leveled by an earthquake in 1887, Charleston disappeared into the mesquite. On the foreground slope *Acacia vernicosa* has registered a sharp increase and now dominates a sparse flora that includes cane cholla, catclaw, desert zinnia, black grama, sideoats grama, and three-awn. Cottonwoods line the river channel, delineating faithfully its backward-S curve. The terrace sides, once grassy, support an almost impenetrable thicket of *Acacia vernicosa*.

PLATE 52*a* (About 1884). With this plate the photographic progression down the San Pedro River is interrupted, and the camera records upland conditions about seven miles east of the river. The view is due north toward the Dragoon Mountains, across Tombstone in its silver age. The Cochise County court-house, just constructed, stands at the far left. The hills on the near side of town present an unmistakable picture of a grassland community. The shrubs dotting it cannot be identified, but probably include agaves, beargrass, and sotol. In the foreground Wright lippia, century plant, and sotol can be recognized. Like Charleston before 1890, Tombstone clearly lay in the grassland. Elevation 4850 ft.

PLATE 52*b* (1960). The grassland has given way to the vegetation of the Chihuahuan Desert. On the hills in front of town creosote bush, ocotillo, tarbush, mesquite, and *Acacia vernicosa* are the principal plants. The foreground has undergone an appreciable invasion by ocotillos, many of which have been cut away to clear the view. On the north-facing slope in the foreground the flora is varied and dense; it includes Wright lippia—the dominant —beargrass, sotol, century plant, tarbush, mariola, mesquite, and a species of *Brickellia*. Because of the poor resolution in the background of the old photograph, it is difficult to generalize about conditions on the bajada behind town. Much the same transition from grassland to scrubland has probably occurred there, however. The region is now dominated by an *Acacia vernicosa*—creosote bush—tarbush community whose character-istic drab hue is familiar to any traveler along Highway 80. Mortonia is also an abundant bajada plant. The arrows point to Walnut Gulch, where the next four pairs of photographs were taken.

PLATE 53a (About 1890). About six miles west of Tombstone, from a station part way up the north terrace of Walnut Gulch, looking southeast across the gulch toward the south terrace and the Three Brothers Hills. The flood plain is open and largely free from brush. The two dark shrubs at the right designated by arrows look like Mexican tea, a plant still abundant in that location. The clumps of grass are probably sacaton. At left the circle encloses what may be a young mesquite; a large individual occupies the spot today. Elevation 4000 ft.

PLATE 53*b* (1960). The floor of the wash has undergone a pronounced shrub invasion by both mesquite and *Acacia vernicosa*. Normally the two plants do not occupy the same habitat, but here they grow side by side, a juxtaposition conveniently illustrated in the foreground where acacia is in the center and mesquite at the right. Since many small plants of both species are present, it is probably safe to assume that neither invasion has attained its maximum development, and each will continue. The south terrace is shrubbier than before and supports the *Acacia vernicosa*—creosote bush—tarbush community typical of so much of the region. One Mexican crucillo (circle) is visible at center on the slopes. A half-dead walnut tree grows at the right (arrow). Out of sight behind the mesquites are burrobrush, desert willow, and *Rhus microphylla*. In the midground, Mexican tea, palmilla, and some clumps of sacaton.

PLATE 54a (About 1890). This photograph, which partly over-laps the preceding one, is taken from the same station, but looks east. Both pictures show the same vegetation. The shrubs at the south base of the little hill in center are young mesquites (see next plate). The abundant dark plant on the plain to the left of the hill may be tarbush, still locally abundant there. Elevation 4000 ft.

PLATE 54*b* (1960). This picture registers the same invasion by mesquite and *Acacia vernicosa* shown in Plate 53. The mesquite has come in most densely at the base of the terrace, along the course of a small tributary that cannot be seen. The conspicuous bunch grass in the fore- and midground is sacaton, *1*. The abundant smaller plants are vine mesquite (a grass) and bull-nettle. Among the distinctive, easily recognized shrubs near the camera are *Acacia vernicosa, 2;* Mexican tea, *3;* some small mesquites, *4;* and chamiso, *5,* which here is heavily grazed and assumes a low, compact shape.

PLATE 55a (About 1890). From a station on the flood plain of Walnut Gulch looking northeast toward the small hill that appears in the preceding plate. Numbers *1* to *3* designate mesquites which carry over into the new picture; *4* is an *Acacia vernicosa* present in both; *5* marks a cane cholla which is downslope from one in the new photograph. Since the last-named species propagates freely from fallen joints, the plant here may be a vegetative ancestor of the newer one. The mesquites are all young—perhaps five to fifteen years of age—a fact that helps date the invasion and suggests that twenty years before this picture was taken the hill may have been almost completely free from brush. Elevation 4000 ft.

PLATE 55*b* (1962). The mesquites have matured, and the hill is badly overgrown with *Acacia vernicosa*. Both species protect clumps of bush muhly, a grass that commonly is found associated with a "mother" plant. Although some co-mingling exists, the two dominant plants have assumed their normal habitat preference, with mesquite occupying the lowlands; acacia, the rockier uplands. Sacaton, chamiso, and cane cholla also grow on the hillside; sacaton and chamiso in the foreground on the floor of the wash. The *Acacia vernicosa* invasion has probably paralleled that of mesquite. Many of the adult acacias in this picture may have been present in the old one as seedlings.

PLATE 56a (About 1890). From an island in the middle of Walnut Gulch about a mile below the site of the preceding plates, and looking southwest down the wash toward its junction (right) with the San Pedro River. The river is inconspicuous here; however, its course is clearly defined in the new photograph by cottonwood trees. The Huachuca Mountains are in the background at the left. As in the preceding plates, the valley floor is open and apparently grassy. The large shrubs are probably mesquite. Elevation 3950 ft.

PLATE 56*b* (1960). From a point twenty-five feet south of the old station, the view from which is obscured by mesquite. The wash now sweeps to the right of the island, having changed its channel. Along the abandoned bed rabbitbrush, desert willow, burrobrush, and *Carlowrightia* (*linearifolia?*) have become established. The north-facing terrace slope at the far left, much brushier than before, has been invaded by Wright lippia, *Acacia vernicosa*, tarbush, and creosote bush, the last named becoming dominant toward the top of the slope. The valley floor is heavily overgrown with mesquite, rabbitbrush, burrobrush, desert willow, and some very large catclaw trees. The water tank in the midground at the left belongs to the railroad station at Fairbank. In front of it and to the right is a small, light-colored hill that presents an interesting anomaly. The slope visible in this picture has apparently changed very little, and supports a heavy cover of black grama, with some bush muhly near the few shrubs. The south-facing side, however—away from the camera—is brushy, boasts many dead mesquites, and is dominated by the typical *Acacia vernicosa*—creosote bush—tarbush community.

PLATE 57a (About 1890). From a hill west of Fairbank looking southeast across the junction of Babocomari Creek (lower arrow) with the San Pedro River (upper arrow). The Mule Mountains are at the right; the Tombstone Hills at the left, with the south terrace of Walnut Gulch, the site of the stations for the preceding four pairs of plates, rising in front of them. This rare photograph, lacking though it is in resolution, verifies much of what has been alleged about conditions along the rivers before the onset of arroyo cutting. Neither stream has a distinct channel. Babocomari Creek winds sluggishly through a marshy, grass-choked plain; and the course of the San Pedro is almost invisible. Except for the mound in the right midground, the site of the modern railroad station at Fairbank, the valley appears to be treeless. Elevation 3800 ft.

PLATE 57*b* (1962). In carving a channel, part of which is visible as it sweeps around the lower left, Babocomari Creek has cut deeply into the hill from which the original picture was taken. The new camera location is about one hundred feet northwest of the old, which no longer exists. The dike at the left of the picture forms part of a project to divert the creek, now threatening Highway 82, to a new channel, visible at the center and about on line with the water in the old picture. The San Pedro's trench is obscured by brush; the junction of the two streams is off-picture to the left. Mesquite, cottonwood, and Goodding willow choke the valley floor. Sacaton is the dominant grass, and forms with mesquite a distinct community at many places along the San Pedro.

PLATE 58*a* (About 1890). From a point about three miles south-east of St. David looking east across a low range of hills toward Cochise Stronghold, center, in the Dragoon Mountains. The San Pedro River is out of sight behind the hills. In 1890, the area evidently lay at the lower edge of the grassland. A few shrubs, many of them palmillas, dot the picture, but by and large the plain is open, uneroded and, although detail is missing, tufted with short grass. Elevation 3700 ft.

PLATE 58*b* (1962). The camera station has been shifted about twenty feet away from the old location in order to avoid a large mesquite. Mesquite, catclaw, and *Acacia vernicosa,* which dominates the foreground, are the most important shrubs on the lowlands. Palmilla (an inflorescence of which appears at right) and cane cholla are less abundant, but still important. Of the five, only the palmilla plays about the same role as it used to. Little remains of the old grassland. The ground, bare and eroded, supports only a scanty understory of fluffgrass, burroweed, and desert zinnia. On the hills, which look about the same, are *Acacia vernicosa* and creosote bush.

PLATE 59a (About 1890). Taken a few feet away from the station for the preceding pair, this picture looks south-southeast across the San Pedro River toward the Tombstone Hills. In this direction the plain appears even more open than in Plate 58. Except for a few yuccas and some unidentifiable shrubs, grass, closely cropped, appears to be the primary cover. Cottonwoods are conspicuously absent from the lowlands along the river. Elevation 3700 ft.

PLATE 59b (1962). All of the plants of Plate 58b occur here and, in addition, tarbush, sacaton, and *Bahia absinthifolia,* which appears as one of the abundant small herbs in the foreground, mixed with desert zinnia, fluffgrass, and burroweed. In the mid-ground a lower, probably moister area has a dense cover of mesquite, *Acacia vernicosa,* sacaton, and bristlegrass. The tall trees in the background are cottonwoods, also recent invaders.

PLATE 60a (About 1890). This picture, taken about one mile southeast of the two preceding plates, and looking due north, is of interest primarily because of the shrubs in the foreground. The one farthest to the right along the lower edge of the photograph is probably chamiso, but the others look like nothing that can be found in the area today. The best guess is that they are mesquites that have been chopped for firewood, and that have resprouted from the stump. Since American settlement in this area dates from around 1879, the chronology for this explanation is about right. Elevation 3700 ft.

PLATE 60*b* (1962). The camera is only approximately at the site of the old picture. Mesquite and gray thorn are the large shrubs visible. Chamiso and one small palmilla (right foreground) make up the remainder of the vegetation. At the left side of the pyramidal hill in the right of the picture a grove of cottonwoods marks the course of the San Pedro River.

CONCLUSIONS

In the past, the plant dynamics of the desert grassland have been studied more intensively than those of the other life zones, and the principal conclusion to be drawn is already well known: that woody species have taken over the dominance formerly exercised by the grasses, which are in decline.

Several of the species in question have been identified in the preceding chapter as invaders of the oak woodland. Mesquite, ocotillo, turpentine bush, desert broom, rabbitbrush, one-seed juniper, cottonwood, and Goodding willow have apparently increased their density in both life zones, although the last two, properly speaking, belong to neither the oak woodland nor the grassland, but to a gallery forest that threads its way through several zones.

The plates show that an increase in some thirteen additional species can be detected at one or more of the stations devoted to the grassland. Some of these plants undoubtedly are invaders of the oak woodland, as well, but either did not happen to be present at the localities sampled there or, if present, could not be quantitatively evaluated by the methods of repeat photography.

Altogether there are twenty-one species that can be categorized as invaders of the desert grassland, some shared with the life zone above, some not. They can be divided into three groups: those noted only in the Santa Rita area (seven species); those only in the grasslands of the San Pedro Valley (nine species); those common to both sectors (five species). In the listing below, the figures denote the number of stations at which an increase was observed.

Invaders observed at the nineteen stations in the San Pedro grasslands. Acacia vernicosa, 15; cottonwood, 7; tarbush, 4(?); Goodding willow, 3; Wright lippia, 2; desert willow, 2; threadleaf groundsel, 2; wait-a-minute, 2; rabbitbrush, 1. Cottonwood, Goodding willow, and rabbitbrush occur not in the grassland proper, but in riparian habitats.

Invaders observed at the fifteen stations in the Santa Rita grasslands. Mortonia (which has probably also increased in the San Pedro), 2; white thorn, 2; turpentine bush, 1; burroweed, 1; Russian thistle, 1; hopbush, 1; one-seed juniper, 1.

Invaders at the thirty-four grassland stations. Mesquite, 29; ocotillo, 10; catclaw, 5(?); gray thorn, 2; desert broom, 2(?).

Invaders from the Chihuahuan Desert. Many plants in the first two categories enjoy wide ranges and no doubt should be placed in the third group also; but for one reason or another, they did not happen to be noted at the stations for the other sector. Conspicuously absent from the Santa Rita listing, for example, are plants of valley and gallery forest. The reason is obvious: washes and valleys were sampled less extensively in the western part of the life zone.

In spite of uncertainty about the extent of overlapping, it is apparent that the eastern group includes one important element lacking in the western: *Acacia vernicosa* and tarbush, which are plants of the Chihuahuan Desert, have been important invaders in the San Pedro Valley.[3] *Acacia vernicosa* has, in fact, come to exercise a dominant position in the vegetation of the uplands and the bajadas within the valley, although before 1900 it evidently played but a minor role. Today it occupies in the eastern uplands the same importance that mesquite has assumed on the bajada around the Santa Rita Range Reserve (Plates 18, 19, 38), and that white thorn has assumed on terraces near Tumacacori Mission (Plates 40 and 41). Together with tarbush and creosote bush, it forms a distinctive community on limestone soils. In company with Wright lippia, it inhabits rocky hillsides. By itself it dominates large stretches of gentle bajada, where it may form nearly impenetrable thickets in areas that seventy-five years ago were grassy and open. Its proliferation must rank as one of the major events in recent vegetative developments in the grassland.

VI

THE DESERT

Judged either by the diversity of life forms to be found in it, or by the number of species making up its flora, the Sonoran Desert supports the most complex vegetation of the four arid regions of North America. At its most varied, in the upland situations of southern Arizona where broken terrain produces a variety of microclimates and where coarse, rocky soils in conjunction with well developed drainage patterns give rise to similarly heterogeneous edaphic environments, a single vicinity may contain plants representing more than half of the twenty-five life forms suggested by Shreve (1964) as a basis for classifying desert plant life. Throughout the mosaic the prevailing stress may be that of moisture deficiency. But the ways are many by which species have adapted to this preeminent condition through modification to anatomy, physiology, or morphology. And although a single community, occupying a single patch of the mosaic, may include only a few species, there are so many communities that the total number of plants growing in the Sonoran Desert is large indeed.

On the basis of characteristics shared by the dominant members of the vegetation Shreve (1964) has subdivided the desert proper into seven regions that can be conveniently referred to as provinces. In general, broad climatic and topographic factors shape the provincial boundaries, but within the bounds of a single province the vegetation may range from simple to complex depending on variations in soil and microclimate. Descending the gentle outwash slope, or bajada, which surrounds a desert mountain range, one finds a progressive change toward simplicity as the terrain becomes smoother, as the soil tends to become finer and more uniform, and as the gradual loss of a well defined drainage pattern, all tend to produce homogeneous soil conditions. Between the upper part of the bajada and the lower, many perennial species drop out; only a few new ones enter. The life forms become fewer in number, with a marked tendency in the lowlands for those evergreen plants to become dominant that are capable of biseasonal growth.

Some of the effects of microclimate have already been discussed. The importance of soil in determining the character of desert vegetation resides in the strong control it exercises over moisture through its regulation both of the quantity of water available and the duration over which moisture is present (Shreve 1964: 21). Although much of the work in the North American deserts has stressed the importance of salt content, Shreve, himself, made little of this soil factor, stressing instead such physical features as texture, depth, and surface characteristics.

Yang and Lowe (1956) have shown that the relatively level sites at the bases of bajadas are more xeric than the slopes themselves because of soil texture differences.

The effect of soil depth may be seen in the reduction of perennial plant cover to values of five per cent or less where the relatively coarse, absorbent, upper layers of soil are shallowly underlain by impervious horizons of caliche or of hardpan. Here, the downward movement of soil moisture is impeded, and much of the water that might otherwise be available to plants is lost by evaporation.

Surface configuration may influence plant life either by inducing heterogeneous vegetation where the soil surface is dissected and irregular, or by inducing relatively simple plant cover where level terrain, undissected by runnels, imposes more uniform soil moisture conditions. Perhaps nowhere in the Sonoran Desert are differences so stark between the plant life of these two kinds of terrain as they are in the first of the major provinces, the Arizona Uplands.

THE ARIZONA UPLANDS

By reason of the relative abundance of moisture they receive and the wide range of elevations they span, the Arizona Uplands are the most diverse of the three provinces of the Sonoran Desert dealt with. They lie along the northeastern edge of the desert region (Figure 10), where true desert vegetation may extend upward to elevations above 3000 ft. on warm, south-facing slopes. From that approximate upper limit, the plant life of the province extends downward toward the west and south to elevations of about 1000 ft. Average annual rainfall varies from 7 in. to 12 in., the amount being closely dependent on elevation (Green 1959). Mean annual temperatures range roughly from 64° F. to 72° F.

A bajada may once again be used to illustrate the

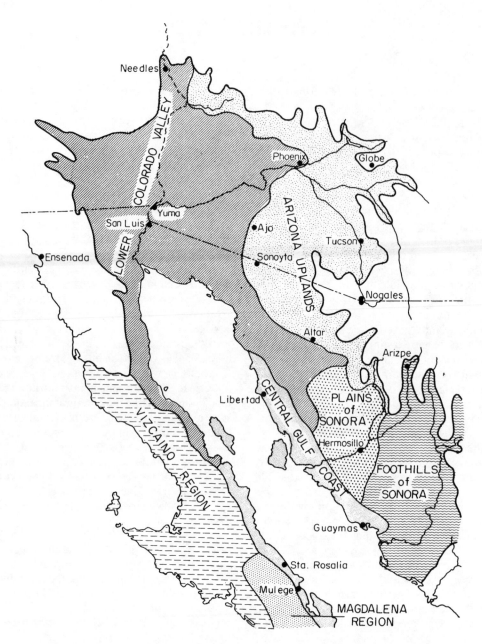

FIG. 10—The vegetative provinces of the Sonoran Desert (After Shreve 1964).

variety of vegetation to be found. On the higher reaches of the slope 40 per cent or more of the surface may be covered by the crowns of woody and succulent perennials. Here desert growth reaches its most luxuriant stage, often producing a cover through which it is difficult to see for more than a few hundred feet. Beneath the low shrubs, sometimes occurring with marked fidelity, one finds such small plants as pincushion cactus. Above these, bursage commonly grows, in a uniform layer about one foot high. Their crowns perhaps twenty feet higher still, low trees like foothill paloverde and ironwood occur between depressions in the drainage. And towering over all the others, widely spaced saguaros may often reach heights of thirty to forty feet. Between the two clear strata of the low trees and the bursage, there are many other plants. They vary so greatly in stature, however, that no layer of intermediate height can be recognized. The entire complex has been referred to as the paloverde-saguaro association by Yang and Lowe (1956), as the paloverde, bursage and cacti desert by Nichol (1952), and as the paloverde—triangle bursage range by Humphrey (1960).

Although several plant species occur together, each may respond uniquely to seasonal changes in moisure and temperature, so that there is no clear rhythm of activity within the bajada community except at the onset of the summer rains, when moisture becomes suddenly available at a time when temperatures are favorable for growth, thus breaking the long aestivation of foresummer. Several species, like limber bush, leaf out only with this coincidence of available moisture and high temperature. Another group, including bursage and creosote bush, produces leaves during all seasons when there is water. Intermediate between these two extremes are white thorn, ocotillo, and foothill paloverde which do not produce leaves during the cold winter, but may foliate in the spring, summer, or autumn if soil moisture is present in sufficient quantity. In still another phenologic category, the saguaro is able to absorb and accumulate water during the winter and summer, but no growth occurs until the warmer spring and summer periods.

The variation in the vegetative activity and the lack of synchronization among the species involved are both expressions of the diverse manner in which plants may fill the many ecologic niches in the heterogenous upper bajada. When one considers leaf fall and flowering habits, the diversity becomes still more striking. Ocotillo loses its leaves abruptly with the onset of drought; foothill paloverde retains its foliage over longer periods of dryness. Bursage loses its leaves slowly, the older ones first and the younger ones later, until with prolonged drought only the very youngest leaves, at the stem tips, remain; these too may die if desiccation is extreme.

Flowering is often less dependent upon rainfall than is vegetative growth. Particularly among the cacti, flowers may appear without regard to moisture conditions; thus, the saguaro blooms unfailingly each year, the flowers usually opening during May in advance of the summer rains. Foothill paloverde does not bloom each year, and its failure to do so is apparently related to insufficient winter rainfall; yet, in years when floral production does occur, flowering comes in May during the arid foresummer (Shreve 1964, Turner 1963). Other plants, such as creosote bush, may produce flowers during any month of the year when moisture is adequate.

A second conspicuous community of the bajadas occurs in narrow, branching ribbons along the drainage channels that interrupt the paloverde-saguaro community, and is analogous to the gallery forests of higher elevations. White thorn and catclaw are most typical here, their size depending upon the amount of moisture available. Where minor washes coalesce in a single large channel, these plants reach heights of 15 to 20 ft. and may be joined by mesquite, blue paloverde, desert willow, canyon ragweed, and other species requiring the improved moisture balance of this habitat. Several of these desert riparian species are among those plants known to have invaded nonriparian positions above the Sonoran Desert in the desert grassland and oak woodland.

As the base of a bajada is approached, many perennials of the upper slope are no longer present; ironwood, foothill paloverde, bursage, and saguaro, among others, are gradually lost in the descent. Toward their lower limit, the indivduals of these species become confined to small drainage ways, with the intervening areas now occupied by creosote bush and white bursage, members of the community from the plain below.

This third community is low in stature and simple in composition, two features which bespeak the less favorable moisture supply to be found on the basal plain. Creosote bush is the principal dominant, occurring as a widely and rather uniformly spaced plant about two feet high in the drier habitats and over six feet high where moisture is more abundant. Under optimum conditions it may attain coverage values of 15 to 20 per cent. White bursage, its principal associate, occupies the broad openings among the larger shrubs. It rarely exceeds two feet in height and may have coverage values of from less than 1 per cent to 10 per cent. Total coverage for the community may vary from 15 to 30 per cent, depending on the availability of soil moisture. Unlike the paloverde-saguaro association, the creosote bush—white bursage community is not restricted to the Sonoran Desert, but occurs along the valleys of the Mohave Desert as well (Shreve 1964, Allred et al. 1963).

A similar distribution characterizes still a fourth

plant community of wide extent in the Uplands. Although restricted in the Sonoran Desert to the northern part, areas dominated by desert saltbush, like those dominated by creosote bush and white bursage, also extend into the Mohave Desert. It occurs in essentially pure stands on a variety of soils: on well drained, fine, sandy loam; on other types where surface drainage is impeded (Shantz and Piemeisel 1924). The total plant coverage expresses the relative water balance of each location; coverage values as low as 3.5 per cent have been noted in a site where downward percolation is blocked by hardpan. As with many plants of the desert plains, vegetative growth occurs during both rainy seasons; flowering is restricted to autumn. According to Aldous and Shantz (1924), the dominant vegetation of the bottom lands along the Gila River is also desert saltbush. Thus the association extends from the Arizona Uplands through the second of the provinces, the Lower Colorado Valley.

THE LOWER COLORADO VALLEY

Proceeding westward along a line connecting Tucson, Arizona, with San Luis, on the Colorado River in Mexico, one crosses a series of plains that descend to the river like giant stairs, each separated from the other by mountain ridges that grow lower and less massive the farther one travels from the backbone of the continent. With each tread the plant life grows more impoverished; the simple creosote bush—white bursage community is no longer confined to the valleys; where foothill paloverde, bursage, ironwood and saguaro do occur, they, in contrast, are restricted almost entirely to drainage ways. Beginning about at Ajo the vegetation of the Arizona Uplands grades into that of the Lower Colorado Valley.

The final 250 mi. of the Colorado River flow through this low, arid province, which extends eastward along the course of the lower Gila for another 200 mi. To the west it reaches 150 mi. into the Imperial Valley of California; to the south, narrow extensions into Mexico flank the head of the Gulf of California.

Throughout its extent, the province is confined mainly to elevations lower than 2000 ft., and in the Imperial Valley it extends down to well below sea level. It is at once the hottest and the driest of the desert subdivisions. Mean annual temperatures for stations within it range from 65° F. to 74° F.; the average precipitation from slightly more than 1 in. per year to slightly less than 8 (Green and Sellers 1964; Hastings 1964a, 1964b). Biseasonal in distribution, the rainfall tends to be equally distributed between winter and summer or, in the westernmost reaches, slightly unbalanced in favor of winter.

As one might expect from the degree of aridity, the vegetation is simple, sparse, and relatively uniform.

The commonest community is one dominated by creosote bush and white bursage. Here, in contrast to their distribution in the Arizona Uplands, these two species are not confined to valleys. They cover vast stretches of plains, bajadas, and even volcanic hills, their density and height varying with the amount of moisture available from the local soil. On some volcanic outcrops a depauperate phase of the paloverde-saguaro community of the Arizona Uplands may be present instead. Under these conditions, foothill paloverdes characteristically are low and widely spaced; and as one approaches the Colorado River, the infrequent saguaros disappear almost completely.

The runnel vegetation of the province includes foothill paloverde and ironwood, but lacks white thorn, one of the more conspicuous plants in similar habitats in the Uplands. Along some of the broader washes mesquite and blue paloverde occur. The smoketree, absent from most of the Arizona Uplands because of its sensitivity to cold, grows here in riparian situations (Benson and Darrow 1954: 208); conversely desert willow is absent below about 1500 ft. (Benson and Darrow 1954: 313), although it is abundant along washes in the province to the east, and in the grasslands.

Expanses of stabilized sand characterize large areas of the Lower Colorado Valley, and the typical community on these is dominated by big galleta, growing either in pure stands or in company with creosote bush and other perennials (Plates 82 and 83). The apparent anomaly of a grass-dominated community existing under a climate favorable to shrubs and low trees may be best understood through a knowledge of this plant's life form. Unlike most grasses, big galleta has perennating buds borne above the soil surface on woody culms, a characteristic shared with many desert shrubs. Although its phyletic relationship is with the grasses, areas in which it is the dominant plant are not analogous to grasslands.

In its physiography the Lower Colorado Valley is more uniform than the other two provinces considered; only a few mountain ranges, and they relatively low, interrupt the plains. Sand dunes and malpais fields—the latter the product of recent volcanic activity—occur toward the south and support distinctive communities of their own. The Pinacate Mountain region (Plates 82–88), where lava flow and dune exist side by side, is a checkerboard of dark and light localities that support an intricate mosaic of vegetation responding sharply to the shifting patterns of soil and albedo. Pockets of relatively deep sand support big galleta; other sandy areas are dominated by ocotillo, and in some cases, even by saguaro; the volcanic hills maintain a sparse cover of foothill paloverde, ironwood, elephant tree, and *Jatropha cuneata*. The latter two plants have affinities to the south where, with their fleshy-stemmed relatives, they form the characteristic vegetation of the third and last of the provinces dealt with, the Central Gulf Coast.

THE CENTRAL GULF COAST

The vegetation of the Central Gulf Coast occurs in two coastal strips along opposite sides of the Gulf of California; to the north both give way to the Lower Colorado Valley and today, if not in the past, the two strips make no contact with each other. The province has been described by Shreve (1964) as the driest in the Sonoran Desert, but from the weather records available now this distinction clearly belongs to the Lower Colorado Valley (Hastings 1964a and 1964b). The coastal region seems to be characterized more than anything else by rapidly changing gradients of rainfall. The isohyets are closely crowded, more or less paralleling the coastline, and the amounts fall precipitously as one approaches the Gulf from the east. Guaymas, one of the few stations with a long-term record, receives an annual average of about 9.5 in., with 75 per cent falling during the six hot months (Hastings 1964b). This amount is probably near maximum for the province; most parts receive between 4 and 6 in.

In contrast to the two provinces already described, the vegetation of the Central Gulf Coast lacks the sharp distinction between communities of the plains on the one hand and the bajadas and hills on the other. As the traveler approaches the region from the Lower Colorado Valley, creosote bush and white bursage become sporadic in their occurrence, and no longer dominate extensive areas. Hills and intermontane plains, valleys, and bajadas alike are covered by open communities of arborescent or shrubby physiognomy.

Relative to the Arizona Uplands, foothill paloverde loses its dominant position in the vegetation, and ironwood and blue paloverde gain in importance. Elephant tree and *Jatropha cuneata*, plants of limited occurrence in the Lower Colorado Valley, are joined by a host of other fleshy-stemmed, sometimes aromatic plants: *Bursera hindsiana, Jatropha cinerea, Euphorbia misera*. In most situations, none of the perennials asserts dominance to the degree that foothill paloverde, for example, does at places in the

Arizona Uplands. The saguaro is of minor importance, its place in the vegetation being occupied now by the *cardón,* or by other columnar cacti like the organpipe cactus and *sinita.*

The boojum tree, another fleshy-stemmed species (Plates 92 and 94) is found at only one location on the Mexican mainland, but occurs abundantly there. Its maximum age, judging by its growth in the interval between pictures taken in 1931–32 and 1963, must be close to 400 yrs.; thus it ranks among the longest lived of the Sonoran Desert plants for which such information is available. The mainland occurrence of this grotesque tree has been described by Shreve (1964) as covering only a few square miles in the vicinity of Puerto Libertad (Punto Kino or Punto Cirio); Aschmann (1959), however, extends its range nearly as far south as Desemboque, and he is evidently correct. The disjunct distribution of the boojum tree from its present center of occurrence in Baja California to the single, minor outpost on the mainland remains one of the many intriguing problems involving the flora of the Sonoran Desert.

An important community, found also in other provinces of the desert with coastal contacts, is dominated by *Frankenia palmeri* (Plate 95). Often associated with *Atriplex barclayana,* it dominates a low, windswept community that typically is found just inland from the strand. On sandy soils the monotonous appearance of the association may be broken by taller plants found abundantly inland as well: *Jatropha cuneata, J. cinerea, Euphorbia misera,* jojoba, and teddybear cholla (Plate 94).

The photographs that follow sample only two localities within the Central Gulf Coast. Plates 89–95 show the vegetation around Puerto Libertad, Sonora, and, in general, reveal remarkably little change during the past thirty years. Neither the density of most species nor the size of individual plants has appreciably altered. The exceptions to this generalization are found most notably among such columnar plants as *cardón* and the boojum tree, which have increased in size, and, in the case of the *cardón,* in density as well. In the much altered landscape of the Islas Melisas in Guaymas Bay, the second of the two places in the Central Gulf Coast for which photographs are available (Plates 96 and 97), *cardón* is similarly involved.

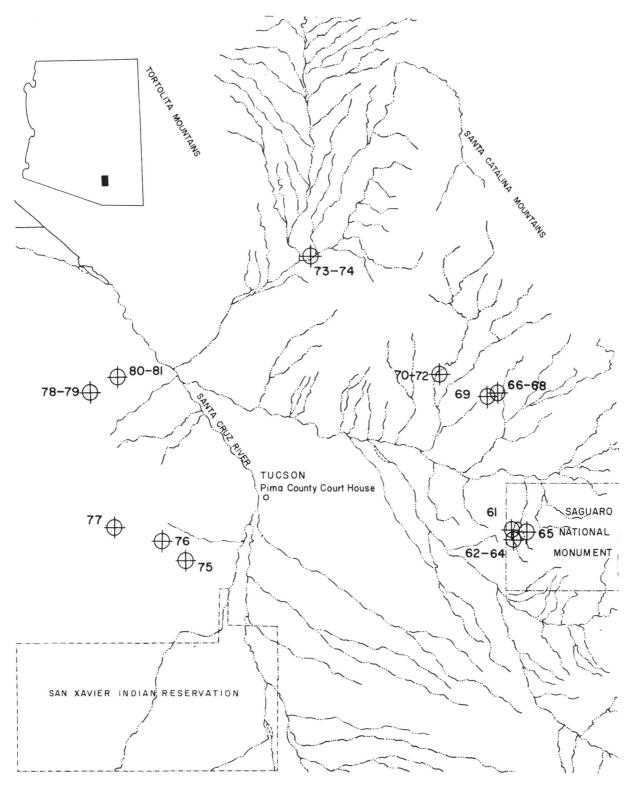

FIG. 11—The location of photographic stations in the Arizona Uplands.

PLATE 61a (About 1935). This highly developed but decadent community introduces the Sonoran Desert proper and its subdivision, the Arizona Uplands. The province is characterized by the association of foothill paloverde and saguaro that appears here in a particularly memorable stand. The camera, located on a hill in Saguaro National Monument near Tucson, faces northeast toward Agua Caliente Hill and looks down a gentle incline that drains into Tanque Verde Creek. Near the bottom of the slope saguaros become infrequent and mesquite predominates. The latter is joined on the stream banks at the right of the picture by larger trees like cottonwood and Goodding willow. Elevation 3050 ft.

PLATE 61*b* (1960). The saguaro population has undergone a reduction by about one-third since the earlier picture. A recent study of the stand indicates that if the present trend continues, the cactus will disappear by 1998 (Alcorn and May 1962: 157).

PLATE 62a (About 1935). This picture, taken about one-half mile south of the preceding one, looks north-northwest toward the Santa Catalina Mountains. Some of the giants are thirty-five feet tall and one and one-half centuries old. The hole in the tall stem at the right has probably been utilized as a bird's nest. Such entries are usually walled off from the main body of the plant by a hard, ligniferous material (Berry and Steelink 1961), but in the case of intrusions by larvae of the *Cactobrosis* moth, the carrier of a bacillus that produces bacterial necrosis, the cactus may fail to respond quickly enough and may become diseased. Elevation 2950 ft.

PLATE 62*b* (1962). Only seven individuals—about one-fourth of the earlier number—still remain standing. Although many of them have died from bacterial necrosis, the disease has probably always been a major cause of death. The stand is not disappearing because old individuals—from whatever cause—die; the failure of young plants to become established in sufficient numbers to replace the old—or indeed, in much of the Saguaro National Monument, to become established at all—is responsible. In the foreground, staghorn cholla, desert hackberry, and creosote bush. The trees are foothill paloverde and mesquite. White thorn, Mexican crucillo, gray thorn, and catclaw constitute the principal shrubs in the midground; desert zinnia, paperdaisy, burroweed, and brittlebush, the more important low perennials.

PLATE 63a (1935). Looking south-southwest at a hillside about one hundred yards from the preceding location. The flora of the wash in front of the hill includes mesquite, catclaw, creosote bush, and desert hackberry. The coarse, gravelly slopes support the usual dominants, foothill paloverde and saguaro; plus two cacti, staghorn cholla and bisnaga; and a variety of shrubs: Mexican crucillo, creosote bush, mesquite, and ocotillo. Elevation 2950 ft.

PLATE 63*b* (1960). The death-strewn hillside illustrates a fact that has been widely overlooked by naturalists preoccupied with the decline of the saguaro: the giant cactus is dying out, but so, in many parts of the Saguaro National Monument, is much of the rest of the community associated with it. In this particular locality the number of saguaros is about 40 per cent less than it was in 1935. But in the stand of chollas marked by arrows in the old picture the decline has been even steeper. The small perennials have also undergone a severe attrition. Of the paloverdes present on the hill in 1962, more than one-fourth were dead. This fraction has, of course, no meaning without reference to the plant's average lifetime, which is several hundred years. It acquires relative significance from a statement made by Shreve in 1911: "I have had a great many thousands of Palo Verdes come under my observation within a radius of 75 miles of Tucson, and have seen only two dead trees of full size" (Shreve 1911a: 296).

PLATE 64a (1935). This picture, taken in the vicinity of the two preceding plates, looks north-northwest toward the Santa Catalina Mountains and provides a good look at some plants that go unrecognized in many of the photographs. Four cacti are present: *1,* saguaro; *2,* staghorn cholla; *3,* jumping cholla; and *4,* bisnaga. Desert zinnia, *5,* is the common low perennial in flower; burroweed, *6,* can be recognized among the desert zinnias. *Franseria confertiflora* occurs as a tiny herb, and *7* may be paperdaisy. Elevation 2950 ft.

PLATE 64*b* (1962). One of the old saguaros, recently fallen, lies rotting on the ground at the right. In twenty-seven years the ranks of the giant cactus have been decimated, and the shrubs—foothill paloverde, white thorn, mesquite, gray thorn, and desert hackberry—have fared almost as badly. Although the number of small perennials is fewer, those here are relatively ephemeral species whose fortunes fluctuate with short-term variations in rainfall. Since 1960, when the area was rephotographed for the first time, the small perennials in general have registered a comeback, and brittle bush in particular, almost extinct in 1960, has made rapid gains. Its behavior here is hard to reconcile with the observations about its apparent stability over a four-year period near the Desert Laboratory (Shreve 1911a: 291).

PLATE 65a (1932). The station is located on a ridge in Saguaro National Monument about one-half mile northeast of that for Plate 61 and faces north-northwest toward the Santa Catalina Mountains. In some respects the floristic variety in the community is more apparent than real. Ninety per cent of the conspicuous plants belong to one of three families, the cacti, the legumes, and the composites; another three families, the elms, the grasses, and the caltrops embrace the remainder. Elevation 2900 ft.

PLATE 65*b* (1962). Small perennials have almost vanished from the landscape. Death among the saguaros has been widespread, and its extent among the shrubs is nearly as great. With some exceptions the conditions shown here prevail throughout the whole lower part of Saguaro National Monument, a section characterized by rolling terrain and soils that are coarse, well drained, and relatively homogeneous. Of the parts of the Monu-ment that lie to the east—in the rocky foothills and along the lower slopes of the Tanque Verde Mountains—the pictures are by no means representative. There the saguaro seems to be re-populating and the plant communities appear to be stable. What-ever the factors may be that are responsible for the widespread mortality in the lower Monument, they affect many species, not merely one.

PLATE 66*a* (1915). This and the next three sets of photographs were taken around the mouth of Soldier Canyon at the foot of the Santa Catalina Mountains near Tucson, Arizona. Two sets show the vegetation of the rocky lower slopes. In the remaining two the camera points away from the Santa Catalinas and out across the bajada toward Tanque Verde Creek, the principal drainage for the southeast side of the massif. There are marked differences in the vegetation of the two habitats—mountainside and bajada—and marked differences also in the stability of their respective communities. This view looks north-northeast into the canyon toward the mountains. Elevation 2950 ft.

PLATE 66b (1960). A few teddybear chollas have disappeared; the ocotillo at the left has added some stems; and the small saguaro in the foreground at left center has grown to a respectable size. The Mt. Lemmon highway has taken its toll of the native vegetation but, all in all, there has been little significant change. Foothill paloverde, mesquite, and *Lycium berlandieri* comprise the basic tree and shrub element; brittle bush is the principal semishrub.

PLATE 67a (1915). The camera has swung outward from the preceding view and now looks across the bajada toward Saguaro National Monument, located on the outwash on the other side of Tanque Verde Creek. The view is south-southeast; the Santa Rita Mountains are at the right and the Tanque Verdes at the left. Compared to its density in the preceding plate, the vegetation here looks relatively sparse. By 1960, of the score or more of saguaros in the midground, only the one marked by an arrow still stood. By 1962 none remained. Elevation 2900 ft.

PLATE 67*b* (1962). In contrast to conditions on the mountainside, the vegetation here has been substantially altered through the years. The most obvious change has occurred in the saguaro population. The area appears also to be brushier, although it is not clear whether this is because there are more individuals or merely because their average size is larger. Many relicts occur between the two photographs: *1*, desert hackberry; *2*, mesquite; *3*, foothill paloverde. The Engelmann prickly pear, *4*, in the foreground has come in recently. Also visible are *5*, brittle bush; *6*, blue paloverde; *7*, *Lycium berlandieri; 8*, catclaw.

PLATE 68a (1908). The camera station is seven years in time, but only a few yards in space away from that in Plate 67. Here, however, the view again is northeast—away from the plain and toward the rocky slopes of the Santa Catalina Mountains. Brittle bush blankets the foreground. Saguaro, ocotillo, mesquite, and foothill paloverde can also be identified. Elevation 2900 ft.

PLATE 68*b* (1960). In contrast to conditions on the bajada (Plate 67), the saguaro population here, one hundred yards away in the foothills, has been stable for half a century. The decrease in the amount of brittle bush in the foreground may or may not be significant (see Plate 64*b*). In addition to the plants that can be identified in the 1908 view, catclaw, staghorn cholla, mesquite, and Engelmann prickly pear are present.

PLATE 69a (1908). Here the camera is actually located on the bajada and is about one-fourth to one-half mile southwest of the three preceding stations. It faces east toward the Santa Catalina Mountains (left) and Agua Caliente Hill (far right). Individuals that carry over into the recent view are 1, foothill paloverde; 2, three mesquite trees, two of which are already mature adults; 3, a desert hackberry; and 4, two saguaros. The lower story is relatively dense and, judging from its composition in 1962, contains burroweed, paperdaisy, Mexican tea, and brittle bush. Elevation 2850 ft.

PLATE 69b (1962). Like Plate 67, dealing with the vegetation of the upper bajada, this one indicates a decline in the number of saguaros and in the amount of low ground cover; a general increase in the shrubbiness. The inset, taken in 1960, shows the center saguaro before it succumbed to bacterial necrosis. The scar that appears on it can be seen in the 1908 view. In the two years between the inset and the main picture, the low perennials made a substantial recovery. (Compare with Plate 64.)

PLATE 70*a* (1890). The next three sets of plates were taken along a semipermanent stream where water flows several months out of the year and the rest of the time lies near the surface. This is an early view of Sabino Canyon in the Santa Catalina Mountains near Tucson, looking west across the creek toward the entrance to the canyon. Two distinct habitats are juxtaposed: the rocky slopes, supporting a foothill paloverde—saguaro community, and the stream side, boasting a gallery forest the top of which is just visible along the lower edge of the picture. Elevation 2850 ft.

PLATE 70*b* (1962). On the slopes foothill paloverde is everywhere more abundant, and although saguaros are less numerous in a few locations, a count shows that their total number is somewhat greater; on the lower hillside the increase is striking. At right center on the hill above the road teddybear cholla is more conspicuous, but not necessarily more abundant; the fuzziness of the old picture makes comparison difficult. Mesquite has increased along the runnel that parallels the highway on the near side. Fairyduster, catclaw, limber bush, cockroach plant, brittle bush, hopbush, and *Lycium berlandieri* occupy spaces among the dominants. On the lower hillside, just above the gallery forest: desert hackberry and *Coursetia microphylla*.

PLATE 71a (1890). From the same station as the preceding plate, looking north-northeast
up Sabino Canyon. The cool, moist floor supports a tree growth that contrasts markedly
with the desert vegetation on the slopes. Along the stream, oaks extend downward 1500 ft.
below their accustomed range in the woodland (Shreve 1915: 21). An Emory oak, *1*,
and a Mexican blue oak, *2*, can be identified by virtue of their having persisted into the
present. Many of the small trees in the near midground look like oaks, but their identification must remain tentative. The distinctive ghost-like skeletons of Arizona sycamore,
leafless for the winter, are readily recognized. One of them, *3*, is still alive. Compared
to present-day conditions, the canyon floor is quite open. Elevation 2850 ft.

PLATE 71*b* (1962). An increase in mesquite, cottonwood, and Goodding willow accounts for the cluttered appearance of the canyon. Arizona sycamore, velvet ash, and the two species of oak grow as best they can among the invaders; catclaw, foothill paloverde, and seep willow form secondary components of the gallery forest. On the rocky hillsides foothill paloverde has proliferated. The saguaro population also is greater; as in Plates 70 and 72 the biggest increase has been registered on the lower slopes and on the canyon floor.

PLATE 72a (1890). From the same location as Plates 70 and 71, looking northeast up Sabino Canyon. The left margin of the picture overlaps Plate 71, and the same Arizona sycamore appears in both views. Oaks, sycamore, and velvet ash dominate the scene. An ecologist describing the locality in 1915 noted the importance of these species, but failed to mention mesquite as a constituent of the gallery forest and stated only that there was an "occasional" cottonwood (Shreve 1915: 24). His description agrees well with the conditions portrayed here. Elevation 2850 ft.

PLATE 72*b* (1962). The dense growth on the canyon floor speaks for itself. A count of the number of saguaros visible on the hill behind the bosque shows that the population is slightly larger today than in 1890. Toward the top of the hill there are fewer individuals; toward the bottom and in the valley, more. Foothill paloverde dominates the hillside along with saguaro. Other species present there include desert hackberry, catclaw, ocotillo, *Lycium berlandieri*, limber bush, and teddybear cholla. On the foreground slope: mesquite, catclaw, and foothill paloverde.

PLATE 73*a* (1913). The next two sets of photographs show a riparian habitat somewhat different from Sabino Canyon. Although this wash, Cañada del Oro, carries surface water only after a heavy rain, infiltration through its sandy bed is rapid enough to insure an adequate supply of groundwater at all times for perennials with moderately deep roots. The vegetation correspondingly is more xeric than that on the floor of Sabino Canyon, but more mesic than that of the surrounding desert. The camera looks east-southeast toward Pusch Ridge on the west side of the Santa Catalina Mountains near Tucson; its view ranges vertically from the desert through the oaks to the pine forest, a dark outpost of which is visible on the peak at right center. Elevation 2650 ft.

PLATE 73b (1962). Burrobrush, crownbeard, and a species of stickleaf occupy the foreground and the channel. Mesquite, desert willow, catclaw, white thorn, and pencil cholla comprise the main elements of the thicket on the opposite bank, which is denser than in 1913; an increase in mesquite apparently is re-sponsible. On south-facing slopes in the hills, foothill paloverde and saguaro act as dominants for an elaborate retinue of sub-ordinate plants. The paloverde, at least, has undergone a substan-tial increase in numbers.

PLATE 74a (1913). Looking southeast from the station for the preceding plates. The left margin overlaps slightly with that view, and the composition of the bank and channel vegetation is continuous between the two photographs. Paloverde appears to be less important than it becomes by 1962. Its apparent increase with time is confirmed by an analysis of mortality and relative age among the plants on a study plot located on the hill in the middle of Plate 73. Among 108 paloverdes of both species, there are no dead ones. Of the 30 foothill paloverdes present, 53 per cent are under 3 ft. in height. The balance between establishment and mortality is obviously in favor of the former. These data for a young, vigorously repopulating stand should be compared with those for the decadent populations discussed in Plates 63, 82, and 86; the apparently stable population in Plate 81. Elevation 2650 ft.

PLATE 74b (1962). The flora of these north-facing hillsides, relatively moist and relatively high for the desert, has many elements in common with the desert grassland. In general, three communities can be distinguished: on steeper slopes there are only a few trees; desert hackberry, catclaw, and white thorn are the dominant shrubs; sotol, desert honeysuckle, fairyduster, wait-a-minute, hopbush, burroweed, turpentine bush, and species of *Opuntia*, the secondary plants. Along the gentler gradients that prevail over most of the hillside these same species are to be found, plus ocotillo, but they are subordinate to a tree component made up of blue and foothill paloverde. On warm, dry, south-facing slopes, many of the species typical of the grassland drop out; blue paloverde becomes rare; saguaro enters and asserts its codominance with foothill paloverde.

PLATE 75a (1915). In the Tucson Mountains looking northwest across Robles Pass toward the southeast face of Cat Mountain. The old road is lower than the present highway and follows a different alignment, but the similarity is great enough to be misleading. Like the rest of the Arizona Uplands the area is floristically varied. In 1962 the foreground, a rocky north-facing hillside, supported foothill paloverde, white thorn, gray thorn, and *Lycium berlandieri* as dominants; and an abundant understory of bursage, white ratany, range ratany, fairyduster, cockroach plant, paperdaisy, cholla, Engelmann prickly pear, tobosa, *Janusia gracilis, Opuntia phaeacantha, Ayenia pusilla, Hibiscus coulteri,* and *H. denudatus.* Elevation 2650 ft.

PLATE 75b (1962). In addition to Area I, the hillside, which has already been described, five other habitats can be distinguished. Area II on the near side of the pass and sloping northward, is dominated by foothill paloverde with a light sprinkling of saguaro, bursage, creosote bush, and mesquite. Area III, along a runnel, has no foothill paloverde and its cover is predominantly creosote bush and white thorn. Area IV, on the opposite side of the pass and sloping gently to the south, is similar to Area II, but as might be expected of a warmer exposure, has, in addition to the paloverde, an abundance of ironwood and saguaro. The dominant small shrub is bursage. In Area V, the rocky lower slope of the mountain, brittle bush replaces bursage in the understory, and ironwood drops out to leave foothill paloverde and saguaro as the large dominants. Area VI, above the escarpment line, finds the latter two diminishing in importance and a group of unidentified shrubs, probably including jojoba, taking over. The most important change between 1915 and 1962 is the great increase in the number of foothill paloverde, a change that is most evident in Areas II and IV.

PLATE 76a (1914). The camera station is located at the foot of Cat Mountain about one and one-half miles northwest of the station for Plate 75. The view is south-southeast toward Beehive Peak. Floristically the scene is similar to Area IV in the pre- ceding plate. The abundant foreground shrub is bursage. Foothill paloverde and saguaro are the chief large dominants, but ocotillo, staghorn cholla, and several other species of cacti appear. Elevation 2750 ft.

PLATE 76*b* (1962). Bursage is less abundant in the foreground— a change which may or may not be significant since picnickers have disturbed the area considerably—and jojoba actively competes with it in numbers. Cholla, white thorn, creosote bush, staghorn cholla, and prickly pear also occur. The most notable change is the extent to which foothill paloverde has increased. As a result, the view is much less open.

PLATE 77a (1915). The third shot in a circuit of the Tucson Mountains, this is from a station three miles west of the preceding one and looks west-northwest toward Brown Mountain. Here, as in Plates 75 and 76, the most notable difference over the years is the extent to which the view has closed in. Again an increase in the number of foothill paloverdes seems to be responsible. The arrows mark two saguaros that carry over into the new photograph and that can be readily identified. Elevation 2650 ft.

PLATE 77*b* (1962). The buildings of Old Tucson, constructed since 1915, are visible at the far right. Foothill paloverde (denser than before), and the giant cactus (less dense) are the dominant plants. Other species are *1,* creosote bush; *2,* ocotillo; *3,* staghorn cholla; *4,* Engelmann prickly pear; *5,* limber bush; *6,* white thorn; *7,* bush muhly; *8, Opuntia phaeacantha.* Present but not labeled are fairyduster, catclaw, some species of grama grass, *Coldenia canescens,* and *Hibiscus denudatus.* The common foreground shrub is bursage.

PLATE 78*a* (1916). Looking north-northwest toward Safford Peak (extreme left) from a ridge on the west side of the Tucson Mountains. Saguaro, ironwood, foothill paloverde, and ocotillo dominate the midground; brittle bush the foreground. Among the smaller plants present on the site today are white thorn, creosote bush, white ratany, bursage, senna, staghorn cholla, jumping cholla, *Opuntia phaeacantha, Dyssodia porophylloides, Lycium berlandieri,* and *Janusia gracilis.* Elevation 2900 ft.

PLATE 78*b* (1962). The brittle bush in the foreground has undergone a substantial reduction in numbers, and at least two of the large dominants, ironwood and foothill paloverde, appear to be fewer. A carcass count confirms the visual impression: within a radius of one hundred feet of the camera, 20 per cent of the foothill paloverdes are dead and 40 per cent of the ironwoods. Although mortality like this is abnormally high for these long-lived species, the condition seems to be local.

PLATE 79a (1916). A few hundred feet northeast of the station for Plate 78 and looking north-northwest toward Safford Peak, left center. The hill in front of the camera is at the far right in the preceding plate and, in spite of the different angle, some of the plants can be matched. This picture shows the Arizona Uplands in an advanced stage of development. Foothill paloverde and ironwood constitute the principal arborescent element. Mesquite, omnipresent in the valleys, plays a minor part in this hilly country; only a few individuals occur, and they along runnels. By virtue of their size and shape, saguaros and ocotillos assume a visual importance out of all proportion to their actual numbers. Elevation 2900 ft.

PLATE 79*b* (1962). The foothill paloverde on the right and one ocotillo on the left still frame the view; the foreground looks as cluttered as before. The first hill appears somewhat more open; it has as many saguaros as in 1916, but fewer paloverde trees. Part of the difference in the appearance of the latter's population is real; part may be due merely to the degree of foliation. The paloverde evades the arid fore- and after-summer by remaining leafless; at such times green stems and branches can maintain enough photosynthetic activity to meet the minimum survival needs. In the earlier picture leaves are present; in this one, a November view, they are absent.

PLATE 80a (1908). Looking northwest at a rocky cliff near Picture Rocks in the Tucson Mountains, a few miles northeast of the preceding station. This photograph, by D. T. MacDougal, forms a panorama with the next plate, and the two record some of the floristic detail of the uplands. Foothill paloverde and saguaro dominate the scene; jojoba and creosote bush are important secondary shrubs; limber bush, brittle bush, bursage, and white ratany constitute the dense lower story; and four cacti—teddybear cholla, staghorn cholla, jumping cholla, and bisnaga—lend some diversity to the landscape. Tanglehead and *Janusia gracilis* complete the list of the major species. Elevation 2400 ft.

PLATE 80*b* (1962). A high degree of stability characterizes the locality. Over the entire old panorama one can count 36 saguaros. Of these, 23 survive the intervening years and appear also in the recent shots. Although 13 die, 14 new ones appear, and the visible population remains almost exactly the same at the end of 54 yrs. These figures yield a mortality rate of about 0.7 per cent per year, a value that agrees well with data from other sources indicating a lifetime of about 150 yrs. for the plant (Hastings and Alcorn 1961). The saguaro at center has lost its arms; the woody rods forming the structural framework still protrude from the stumps.

PLATE 81*a* (1908). This, the final plate for the Arizona Uplands, looks north-northwest from the same station near Picture Rocks. The young saguaro at the left—perhaps thirty years old—appears also in the preceding view. *Lycium pallidum,* white thorn, and Engelmann prickly pear occur here, but not in Plate 80. In general the steep, rocky, upper slope constitutes one habitat; the more gentle, gravelly, lower slope another. On the upper, brittle bush is the principal small shrub, but gives way to bursage on the lower. In a similar fashion the stubby teddybear cholla yields to the larger jumping cholla, and foothill paloverde to creosote bush. Jojoba seems equally at home in both habitats. Elevation 2400 ft.

PLATE 81*b* (1962). Only the teddybear cholla has radically changed in number over the years, but its increase is quite striking. Creosote bush proves to be the most stable of the species present; at least three-fourths of the individuals visible in this view can also be seen in the older one. Foothill paloverde shows about the same degree of stability as the saguaro (see Plate 80).

In the old panorama fifty plants are present, thirty persist; twenty die; twenty-one new ones appear; fifty-one individuals are present in the new panorama. Calculations yield an annual mortality rate of 0.7 per cent, a value too high to agree well with earlier data (Shreve 1911a: 293), which indicate a lifetime of three or four hundred years for the foothill paloverde.

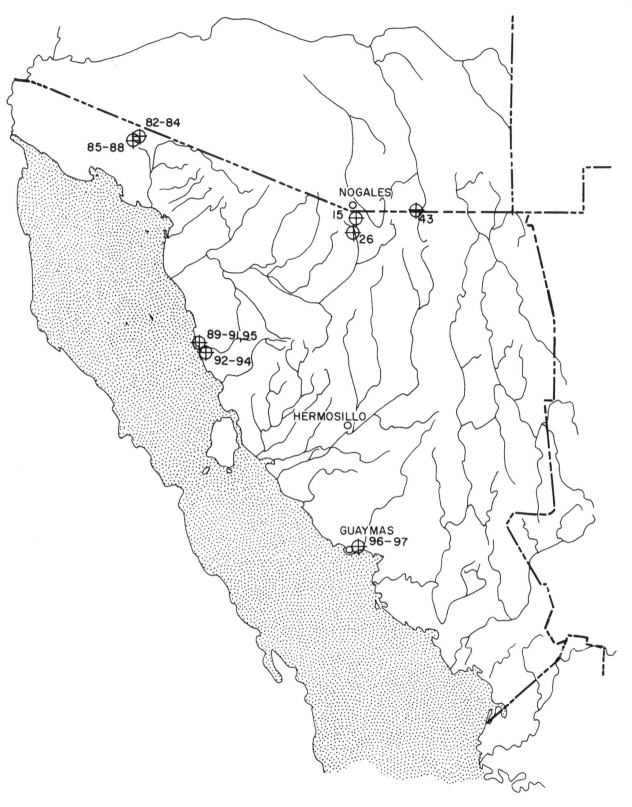

Fig. 12—The location of photographic stations in Mexico.

PLATE 82a (1907). Looking north-northwest toward the United States from a spur of the Hornaday Mountains at MacDougal Pass in the Pinacate Mountain area of Sonora, forty-five miles west of Sonoyta. This hot, sandy plain, flanked partly by volcanic intrusions and partly by dunes drifting inland from the Gulf of California, is near the upper edge of the arid Lower Colorado Valley province, the site of the next six plates. Carry-overs to the new photograph are rare, even among usually long-lived perennials. A positive identification is impossible, but the dominant tree in this picture appears to be blue paloverde, which, like saguaro and creosote bush, seems oddly out of place on a sand flat. Mesquite may or may not be present. Elevation about 900 ft.

PLATE 82*b* (1962). Over a span of fifty-five years, there has been a striking change in the vegetation, not entirely apparent from the picture since dead plants clutter the scene, giving an impression of greater plant density than is really the case. The tree dominants of the earlier view have declined sharply, partly because of the activity of woodcutters operating out of Sonoyta, partly from reasons unknown, but probably climatic. The front two-thirds of the area in this and the adjacent view (Plate 83) have been made into a study plot. Of the 219 mesquite trees, dead and alive on the plot, woodcutters have accounted for 43. Of the remaining 176 mesquites, 91, mostly small, are dead from natural causes; 27 are nearly dead; only 58 of those not tampered with—33 per cent—are alive and healthy. Among the blue paloverdes, a tree that is not suitable for fuel, only one has been chopped. Of the remaining 55, 30 are dead and only 25—45 per cent—alive. Clumps of big galleta—the dominant grass—are also markedly fewer in number.

PLATE 83a (1907). The photograph overlaps with Plate 82 and is from the same camera station looking north-northeast. Three of the saguaros carry over into the new picture, but none of the trees. The crater at left is probably a product of geologically recent volcanic activity in the Pinacate region.

PLATE 83*b* (1962). The high mortality among mesquite and blue paloverde has already been mentioned in connection with Plate 82. Two species of *Lycium*, one of them *L. macrodon*, are present as shrubs, and they have also undergone a substantial reduction in numbers. Of 92 individuals belonging to the genus in the study plot in this and the preceding plate, 28 are carcasses (30 per cent). The increase in saguaro numbers stands in marked contrast to the death prevalent among blue paloverde, mesquite, *Lycium*, and big galleta. Creosote bush is common, although much of it is also dead; bursage and white bursage are both present, although the former is rare and the latter infrequent. A few ironwood trees can be seen.

PLATE 84a. (1907). Looking east-southeast through Macias Pass in the Hornaday Mountains toward the sandy plain shown in Plates 82 and 83. In the foreground, bursage and a large patch of *Opuntia stanlyi*. Beyond, teddybear cholla, creosote bush, ironwood, and saguaro. Elevation about 900 ft.

PLATE 84*b*. (1962). Big galleta now dominates the foreground. Most of the bursage is dead, although many skeletons still stand. The same is true of creosote bush; only two large living clumps can be seen, *1*. The foreground cluster of *Opuntia stanlyi* is still present, but hardly thriving. The density of teddybear cholla seems about the same as before; and about the same number of saguaros appears in each picture, although the individuals are larger in the new photograph and may superficially appear to be more numerous. Two ironwoods, *2*, and a foothill paloverde, *3*, grow on the sandy floor. On the rocky slopes of the Hornaday Mountains: foothill paloverde, elephant tree, and *Jatropha cuneata*.

PLATE 85*a*. (1907). Looking west toward the sands of the *gran desierto*, from the eastern rim of MacDougal Crater in the Pinacate region about 2 mi. from the preceding stations. The crater—a caldera in geological parlance—is about a mile in diameter and 450 ft. deep. Its floor, largely free from the ubiquitous influence of white man, provides a unique opportunity for studying vegetation change. The isolation of the locality, the escarpments ringing the crater, its steep talus sides—all combine to make human entry difficult and, equally important, to eliminate cattle grazing as a factor that has to be considered in accounting for the extensive changes that have taken place. Elevation about 900 ft.

PLATE 85*b*. (1962). Even from a distance of one-half mile the vegetation of the floor presents a strikingly altered appearance. The playa in the center appears desiccated; death is obviously widespread among the plants both there and along the runnels leading into it. Foothill paloverde, saguaro, creosote bush, big galleta, brittle bush, a few mesquites, and a few ironwoods make up the impoverished perennial vegetation of the crater floor. Elephant tree (primarily on north-facing slopes), pigmy cedar, *Jatropha cuneata,* and ocotillo grow on the sides, none of them, however, in abundance.

PLATE 86*a* (1907). An enlargement from the glass plate negative of Plate 85a. Saguaros, foothill paloverdes (the large trees), creosote bush (the small, regularly spaced shrubs), and big galleta comprise most of the vegetation. There are a few iron-woods and mesquites. The talus at the upper left forms part of the crater wall. The apparent foreshortening in this picture relative to the new one stems from the use of a telephoto lens to obtain the latter; the two focal planes are different. Elevation 900 ft.

PLATE 86*b* (1960). The saguaro population is about the same after fifty-three years and, in fact, many individuals appear in both pictures. Among the rest of the vegetation, however, death is widespread; and the scene, in general, is one of arid desola-tion. Only eighteen of the trees in the picture are alive; there are twenty-five carcasses. At least three-fourths of the creosote bush population is dead.

PLATE 87*a* (1962). On the south end of the crater floor, looking northwest across the area viewed from the rim in the preceding plate. The larger carcasses are foothill paloverde; the smaller ones, creosote bush. In this part of the crater floor, marginal to the central playa, creosote bush is living along runnels, but is dead in the interfluvial spaces. The grass, big galleta, is mostly dead. Of the species present, only the saguaro seems to be maintaining itself, and one can expect it to reflect decadence within the next few years. Since its seedlings require shade, the death of the surrounding vegetation ultimately spells decline for it too. A dearth of small saguaros, under a foot in height, is already evident. Elevation about 400 ft.

PLATE 87*b* (1962). An unmatched photograph of a different location in the crater looking across a section of the desiccated playa (see Plate 85). The half-dead mesquite at the left in the midground is the only living tree or shrub in this part of the playa. The hummocks are made up of dead creosote bush, a slow-growing, long-lived shrub that can survive extremely arid conditions, yet here is vanishing.

PLATE 88*a* (1907). From a glass negative by D. T. MacDougal,[1] this photograph forms a near panorama with Plate 85, almost completing a full view of the crater. At center, a volcanic cinder hill; at right, the west end of the granitic Hornaday Mountains. The small, dark shrubs so abundant along the runnels of the crater floor are creosote bush. Elevation 900 ft.

PLATE 88*b* (1962). The plant life is noticeably sparser; the mortality high. Most of the creosote bush on the floor of the crater is dead. In the foreground at the bottom of the picture, one of the rare foothill paloverdes to become established in recent years.

The Central Gulf Coast

PLATE 89*a* (1932). The camera faces west toward hills about four miles northwest of Libertad, Sonora. The common occurrence here of fleshy-stemmed perennials like *1, Jatropha cuneata; 2,* elephant tree; and *3, Euphorbia misera* marks this vegetation as part of Shreve's Central Gulf Coast subdivision of the Sonoran Desert, the setting for the remaining sets of photographs. Other plants visible in the picture are white bursage, bursage, brittle bush, jojoba, creosote bush, ocotillo, and *Frankenia palmeri.* The large columnar cactus is *cardón,* the two multistemmed species in the background, *sinita* and organpipe cactus. Elevation 450 ft.

PLATE 89*b* (1965). On first glance there seems to have been remarkably little change in thirty-three years. Many of the shrubs on the plain below are relicts, and most of these have not changed in size. The ocotillo in the left foreground has apparently produced no new stems since 1932, and the existing stems have not perceptibly elongated. Only the *cardón* has increased in height.

"The slowness of growth, great longevity and low rate of establishment among the perennials gives the desert an extremely stable character" (Shreve 1917a: 216). Despite the general effect of suspended animation among the larger species, a count shows that the smaller shrubs have diminished in number.

PLATE 90*a* (1932). The station is about forty feet away from the preceding one, and the camera looks southeast toward Puerto Libertad, a few buildings of which are visible where the bay curves behind the hills in the midground at the right. In the foreground: ocotillo, brittle bush, *Jatropha cuneata*, and teddybear cholla. Elevation 450 ft.

PLATE 90*b* (1965). These plates, like the pair preceding, seem at first like a study in suspended animation. But here too a close inspection shows that the density and coverage of the smaller perennials have decreased. Among the species visible in the midground are ironwood, foothill paloverde, jojoba, elephant tree, *cardón, sinita, Euphorbia misera,* and a species of *Lycium.* The demarcation between hill and bajada is quite clear, and the relatively rich flora of the former rapidly becomes depauperate and gives way to that of the latter, shown in the next plate.

PLATE 91a (1932). About two miles north-northwest of Puerto Libertad facing west. The picture is taken from a point shown at the left edge of the preceding plate and looks up the bajada toward the small hills at center where the stations for Plates 89 and 90 are located. The poverty of the vegetation of this part of the Central Gulf Coast is evident. Such xeric plants as ocotillo, bursage, and creosote bush occupy the "moist" runnels. *Frankenia palmeri* dominates the ridges between. Elevation 200 ft.

PLATE 91*b* (1965). Thirty-three years later most of the same ocotillos are present and, more surprising yet, many individuals among the smaller perennials have survived. Where lesser plants are this long-lived, changes in their density and coverage take on a significance normally lacking. The later landscape is unmis- takably more arid. Replacement has not kept pace with mortality among bursage, *Jatropha cuneata,* and *Euphorbia misera. Frankenia palmeri* has fared better than the others, but many individuals of the species, which evidently undergoes self-pruning during periods of drought, are smaller than before.

PLATE 92a (1932). On Punto Cirio looking west toward the Gulf of California. In upland situations like this the vegetation of the Central Gulf Coast reaches its maximum development, often displaying a floristic wealth that rivals the Arizona Uplands. Boojum trees, some of them thirty-five feet high and three or four hundred years old, dominate the scene. This and the adjacent region are the only places where this tree occurs outside of Baja California. In the foreground: *1*, elephant tree; *2*, *Jatropha cuneata*; *3*, ocotillo; *4*, white bursage; *5*, brittle bush; *6*, jojoba; *7*, a small *cardón*. Elevation 150 ft.

PLATE 92*b* (1965). The elephant tree at the right has died. The ocotillo has acquired a few more stems. A certain amount of leisurely growth has been enjoyed by the boojum trees, and rather more by the young *cardón* at left center. The two pictures might otherwise have been taken one year apart. Although not all of them are present in this picture, teddybear cholla, desert lavender, creosote bush, *sinita*, bursage, stickweed, *chuparosa, Solanum hindsianum, Bursera hindsiana,* and species of *Ferocactus* and *Lycium* may also be found in the wash.

PLATE 93*a* (1932). Between Punto Cirio and Puerto Libertad, looking southwest across a mesa toward the Gulf of California. *Cardones* dominate the scene. Like saguaro in the Pinacate Mountain region, *cardón* foregoes a rocky habitat in this arid locality and occurs in sand. The middle story consists of elephant tree, *Bursera hindsiana, Jatropha cuneata, J. cinerea,* jojoba, and teddybear cholla. A branch of *Atamisquea emarginata,* known in the United States from only one locality, protrudes into the right edge of the picture. Elevation 100 ft.

PLATE 93*b* (1962). The fleshy-stemmed *Jatrophas* and *Burseras* have changed very little. The grasses are fewer, and two trends noticeable elsewhere in the Central Gulf Coast province are ap-parent: the *cardón* has proliferated; teddybear cholla has de-clined.

PLATE 94a (1932). This near-shore community, located where the canyon shown in Plate 92 debouches into the Gulf, is atypically dense and varied. *Frankenia palmeri* and *Atriplex barclayana* form the low matrix in which *Jatrophas, Burseras, Euphorbia misera,* and teddybear cholla are imbedded. The camera looks southeast toward north-facing slopes on which all of the plants mentioned in Plate 92 are to be found, plus such species as *Fagonia californica, Hibiscus denudatus, Echeveria pulverulenta,* and desert mallow. Elevation 50 ft.

PLATE 94*b* (1965). Although the camera is too far forward to produce an exact match, many relics can be recognized. Here as with a number of the paired photographs of this province, one gets the general impression of a shift toward drier conditions and a more impoverished vegetation. Teddybear cholla has markedly declined; many plants of *Frankenia palmeri* appear to have smaller crowns than before.

PLATE 95a (1932). A few hundred yards north of Puerto Libertad looking northwest toward the mountains in Plate 89. This
picture is perhaps more representative of the near-shore vegetation than the preceding plate. The small, dense shrub with light
flowers is *Frankenia palmeri*. Among it occur: *1*, a scattering
of *Euphorbia misera*; *2*, one brittle bush; *3*, a solitary teddybear
cholla; and *4*, a mesquite. The dense clusters in midground
are made up of mesquite and chamiso. Elevation 50 ft.

PLATE 95*b* (1965). The camera is a few inches too far forward, but many carry-overs can be recognized. The brittle bush and the teddybear cholla have disappeared. The community in general appears sparser and more open, especially on the higher ground between runnels.

PLATE 96*a* (1903). One of the Islas Melisas in the bay at Guaymas, Sonora. The photograph is taken from the southernmost of the two islands and looks northeast toward the outskirts of the city. Here, as at Puerto Libertad and Punto Cirio, the desert reaches down to the water's edge. Columnar cacti are incongruously juxtaposed with mangrove thickets. *Cardón* dominates the scene. *Chuparosa, Euphorbia tomentulosa,* and *Zizyphus sonorensis* occur among the rocks; black or red mangrove in the water. Elevation about mean sea level.

PLATE 96*b* (1961). A striking increase has taken place in the number of *cardones*. The young ones are extremely dense, and if they mature, a near forest will prevail. The same sort of increase has evidently occurred on other islands off the coast. The heaviest stands of the cacti to be noted occur in such habitats. Indirect corroboration of the increase can be had from Shreve (1964: 164), who almost certainly was familiar with the Islas Melisas, yet stated that the only heavy stand of the *cardón* he encountered in Sonora twenty or thirty years ago was located above the salt flats near Empalme. An inspection in 1961 showed the Empalme stand to be not only thinner than this one, but less dense than many of the insular colonies that one must pass on the way to Empalme.

PLATE 97a (1903). The camera, located on the island in Plate 96, looks southwest toward the second of the Islas Melisas, the station for Plate 96. In the background at the left is the peninsula that extends east to Punta Baja and forms the southern perimeter of Guaymas Bay. The *cardón* population is only about one-fourth its size in 1961. Elevation about mean sea level.

PLATE 97*b* (1961). The Islas Melisas, like MacDougal Crater, represent a partially controlled situation, and grazing can be dismissed as an ecological factor. The islands are near enough to the shore to be reached by swimmers, and near enough to Guaymas to ensure that they are, in fact, frequently visited. Nevertheless, they should be among the more stable desert habitats: the temperature and humidity are controlled within unaccustomedly narrow limits by the water; animal interference with the plant life is minimal. A major fluctuation in the population of a long-lived perennial is difficult to account for.

CONCLUSIONS

Local variability of plant populations. Relative to the other two life zones, the desert exhibits vegetative changes that are neither so striking nor so consistent. In some localities shrubs have decreased; in others they have increased. The saguaro is less abundant at Saguaro National Monument today than formerly; in MacDougal Pass the reverse is true. Teddybear cholla thrives at one location; at another it is dying.

The extent is not clear to which one ought to expect random and unsystematic fluctuations in the vegetation of an arid region as a concomitant of the high spatial variability in rainfall. It is possible to argue that in the drier reaches of the desert, some localities, even over a period of many years, will receive more precipitation than adjacent ones, and that the vegetation will reflect this unequal distribution by exhibiting a mosaic of stable, decadent, and thriving patches.

But on the other hand, it is equally possible to argue that the present vegetation must be adapted to wide variation in rainfall, or it could not exist where it now does. Extremes resulting from spatial inequalities, even if continued from one year to the next over a period of many years, and reinforced by temporal variability, should mean little to a long-lived perennial, once it has become established. During the first few years of its existence, before it has developed an adequate root-system or, in the case of the succulents, an adequate capacity for water storage, a plant may, indeed, be vulnerable. But to the saguaro, the paloverdes, the *cardón* and possibly the mesquite —perennials that go on producing seed year after year for periods of well over a century—even the loss of many consecutive crops of plants should matter little. Over their lifetimes local mosaics of abundance and paucity in rainfall surely must average out into smooth mean amounts along well defined gradients.

It is not clear which of these two viewpoints ought to prevail. The problem, then, is which of the changes shown in the plates represent random, local fluctuations; which, if any, indicate general conditions from which one might infer long-term variations in climate. All of the observations that follow have to be qualified in light of this basic uncertainty.[2]

Apparent trends in desert vegetation. In general the small, semiwoody perennials like bursage, brittle bush, and desert zinnia have declined in the period between the old photograph and the new. The change may merely reflect recent variations in rainfall, and may not be significant in a longer view. Bursage, for example, has a lifetime of up to 22 yrs. (Shreve and Hinckley 1937: 475).

Among the long-lived plants, the saguaro, the paloverdes, the *cardón,* and the mesquite show changes that may be significant. Patterns of saguaro repopulation vary with topography. On rocky slopes the species seems stable. In level areas where the soil is homogeneous it has declined. The plates for MacDougal Pass (82 and 83) show an interesting anomaly where, growing in sand, it has increased.

The two paloverdes, blue and foothill, show patterns that are alike. In the upper part of their ranges, they have generally exhibited an increase; in the lower part, a decrease. The two observations can be synthesized in the broader statement that the ranges of the two species appear to have shifted upward. The mortality at MacDougal Crater, near the lower limit of their occurrence, is of particular interest, inasmuch as the area has not been subjected to disturbance by either man or cattle.

The *cardón,* all of the stations for which border the Gulf of California, appears to be increasing. Its proliferation on ungrazed islands near Guaymas is of particular interest.

Mesquite has increased in the Arizona Uplands, as it has in the higher life zones already discussed. At the stations for the Lower Colorado Valley—in harsh, arid habitats near the limit of its tolerance—it evidently has declined. Like the paloverdes, then, its range, except in riparian situations, may be migrating upward, away from the low, hot, dry elevations and into the higher, cooler, moister ones.

VII

THE PATTERNS OF CHANGE

Although no generalization holds for all cases, the most common pattern of change evident in the photographs is the upward displacement of plant ranges along a xeric to mesic gradient, a phenomenon noted in Palestine by Zohary (1962: 130). In some cases the humid upper margin of a species' range has expanded; in others the arid, lower boundary has contracted. In still others, changes have occurred in the mid-reaches of the range and these appear to be linked with microenvironmental characteristics, the decreases often occurring in drier parts and the increases on more humid patches of the environmental mosaic.

Since the spectrum of conditions that a species will tolerate is, broadly speaking, genetically fixed, one must suppose in all of these instances that it is the environment and not the plant's preference that has changed. Expressed in terms of what has happened to the habitat, then, the prevailing pattern has been a shift toward drier or hotter conditions, the plant species migrating upward to where the old, favorable conditions still prevail.

Finally, in a number of cases, there is no apparent pattern at all. From the present evidence, these changes appear to be random.

SPECIES THAT HAVE DECLINED

The desert proper. In the Pinacate Mountains of Sonora, one of the most arid of the areas studied, several important desert plants show decadence. Over the past fifty-five years the first of these, blue paloverde, has declined to the point where more than half of the population in MacDougal Pass today consists of dead trees, their remains persisting undecomposed. Although woodcutters are active in the pass, only one out of thirty-six carcasses examined bore axe marks (Plate 82), and it is clear that some other factor is responsible for the mortality.

A related species, foothill paloverde, has declined in a crater in the Pinacates that is inaccessible to cattle, and that can be entered only with difficulty by man. There, where anthropogenic forces may be discounted as a factor, carcasses outnumber the living individuals. In the face of ordinary drought, the foothill paloverde sheds branches, a phenomenon roughly "analogous to the leaf-fall of the deciduous

trees in the dry months of monsoon climates. . . . When a paloverde seedling becomes large enough to withstand the loss of some of its branches in the most critical portion of the year, its life is safe from all but droughts of the most extreme severity" (Shreve 1911a: 295). During such extreme periods, not just scattered branches of mature trees are lost, but the entire plant body dies. The large number of dead mature plants in the crater points to just such conditions in recent times.

Still a third woody desert perennial, creosote bush, ranks among the most drought resistant species of arid America. In the Sonoran Desert "its adaptability to a wide range of conditions is greater than that of any other desert plant" (Shreve 1964: 157), and it has been known to survive with no rain for as long as a year. That there are extensive areas in the Pinacates where it used to thrive, but where all individuals of the species are now dead (Plate 87b, for example), again emphasizes the severity of recent conditions. The lack of sufficient moisture is unquestionably involved, for where scattered survivors do occur (Plate 87) they are to be found along runnels where moisture conditions are more favorable.

Lycium, mesquite, and the grass, big galleta, also appear to be dying (Plates 82, 83, and 86). Thus, the evidence is clear that stands of five woody perennials and one woody grass are declining in the Pinacate region. With the possible exception of the grass, all are long-lived. The lifetime of foothill paloverde has been estimated at 300–400 yrs. (Shreve 1911a); the longevity of creosote bush probably "greatly exceeds 100 years" (Shreve 1964: 157). A range of from 100 to 400 yrs. seems reasonable as an estimate for all five of the woody perennials. That they are simultaneously undergoing a sharp decline can certainly not be interpreted as random fluctuation in their numbers. And since man is apparently absolved of responsibility for the decadence of at least the two species in the crater, the conclusions seem inescapable, first, that an unusual drying trend is responsible, and secondly that it may be the most severe to befall the region in at least four centuries.

The saguaro is the single prominent species in the

Pinacates to fall outside this pattern. At MacDougal Pass it has increased notably (Plates 82 and 83), and in MacDougal Crater has at least maintained itself (Plates 85 and 86). In spite of the present favorable establishment pattern, there is some evidence suggesting that the saguaro is maintaining only a precarious existence at these two sites. Because seedling saguaros require shade for their establishment (Turner *et al.* 1965), and because the woody plants that provide the shade are disappearing, the current crop of small saguaros may well be the last. According to this holistic view of the community, the disappearance of the saguaro merely lags the disappearance of the shade-providing woody perennials with which it grows and in a few more years the cactus may join the other species in eclipse.

The saguaro figures prominently in changes elsewhere on the Sonoran Desert in areas away from the dry fringe of its distribution. In the Arizona Uplands, a relatively mesic province of the desert, it has declined substantially in relatively level lowland habitats, although in foothill situations it is apparently stable. The difference may again lie with soil moisture factors. Lowland soils are usually of fine texture and contain few large rocks, conditions that contribute to rather uniformly xeric conditions under an arid regime. In the foothills, however, where the soils are heterogeneous and contain a high proportion of rock, conditions are neither so dry nor so uniform. In moist pockets on rocky hillsides, the plant has evidently been able to maintain itself, and its numbers appear essentially static (Plates 66, 68, 70, 71, 72, 79, 80, 81).

But in some localities, other factors may also be at work. In the badly deteriorated stand at Saguaro National Monument near Tucson, cultural effects may supplement those of soil and climate (Niering, Whittaker, and Lowe 1963). To so complex a problem, the techniques of repeat photography are ill-suited, and the final answer must await the completion of other studies. Whatever the cause for its failure to re-establish itself at places in southern Arizona, the saguaro is not alone. In certain of the areas where it is decadent other perennials, like foothill paloverde or white thorn, are also disappearing (Plates 63–65).

The higher zones. In the other life zones, the disappearance of oaks at the border where the woodland formerly made contact with the desert grassland is one of the most striking and consistent examples of decadence to be observed. Almost without exception, the numerous oak carcasses examined show no evidence of being chopped, or burned, or otherwise tampered with by man.

In one of the most interesting cases (Plate 49) the enclave of oaks that existed in 1883 at Gird Dam—evidently the relict of an ancient woodland bridge

from the Mule Mountains across the San Pedro Valley to the Whetstone Mountains—has all but succumbed to the stresses of recent decades. In the earlier photograph, scattered Emory oaks on north-facing slopes marked what was probably the lower border of the woodland. The same boundary today, except for a single living oak at the dam site, is roughly ten miles distant and one thousand feet higher. Significantly enough, to the west of the valley in the Whetstone Mountains, numerous dead and dying oaks occur along the present border.

The conclusion that increased soil aridity is the underlying cause for the upward contraction seems inescapable, and is based on two observations: that the greatest decrease in oak numbers has occurred at the hottest and driest limit of their range, particularly in those places where they formerly grew in isolated microhabitats where topography compensated locally for the prevailing aridity around them; that at higher elevations in the woodland, the oak population appears stable.

At the extreme margin of a plant's range, the local pattern of occurrence may offer valuable clues to the factors that limit its distribution (Daubenmire 1959: 345 f.). And as a corollary, toward the center of a plant's elevational distribution, the range of factor intensities from microenvironment to microenvironment within a single year, and from year to year within a single microenvironment are more likely to fall well within the plant's span of tolerance. Populations, therefore, tend to be more stable in such locations, except over long periods of time, or across sharp differences in habitat.

"The minor fluctuations of climate, which have their minimal and maximal values within periods that are as brief as the normal life of a perennial plant, are registered in the infrequency of every species as it approaches its distributional limit and in the scattered individuals which lie farthest out from the main area of occurrence. The secular changes of climate which have their maximal and minimal points many centuries apart are registered in slight movements of the limits of species, the marginal region of scattered occurrence being, of course, the first affected by such movements" (Shreve 1915: 100).

The evolution of the modern woodland offers no clue as to what factor has produced the aridity and is, therefore, responsible for the recent migration. Like its counterpart in California, the oak woodland of southeastern Arizona has evolved from a Pliocene archetype attuned to a pattern of biseasonal rainfall (Axelrod 1950: 252). But whereas the species that make up the modern California type were segregated under the compulsion of changing climate, and are able today to exist with only winter rain, the Arizona woodland still requires a biseasonal regime. In the spring, the trees complete a lengthy process of defoliation, and in normal years produce new leaves (Phillips 1912). For spring growth, they probably depend upon the precipitation of the preceding winter. Should the winter rains fail, the oaks remain leafless until the summer rainy season. In the event that the summer rains are late, the trees remain bare still longer (Phillips 1912: 6, Shreve 1942b: 195), thus nearing the brink of their drought resistance.

The woodland in the desert region, then, probably depends on both winter and summer precipitation. The disappearance of the oaks could be related to a decrease in the amount of effective precipitation during either season or both seasons. An increase in temperature might accomplish the same end by hastening evaporation. So might a change in any one of many factors in the heat and water balance: the albedo of the land surface, the cloud cover, the net radiation; the rate of infiltration; the intensity of rainfall. Soil aridity need not necessarily be equated with a lack of rain, although certainly the first and most obvious link is with rainfall amount.

SPECIES THAT HAVE INCREASED

Three species whose decline has already been noted in the Pinacates at the low, arid margin of their distribution, have, with some exceptions, increased at localities within the Arizona Uplands. In the case of two, foothill paloverde and blue paloverde, the increase has been near the humid, upper limit of their range. The third, mesquite, has proliferated primarily at still higher elevation in the desert grassland and the oak woodland.

At three sites within the Tucson Mountains, there are substantially more foothill paloverdes than there were fifty years ago (Plates 75–77). At one site, however, there has been a general decline, in which the paloverde has participated along with other species. At three other localities in the vicinity of Tucson, the reproduction of foothill paloverde has outstripped mortality (Plates 70, 73, and 74). But in Saguaro National Monument, the reverse is true (Plates 63 and 64), and once again there has been a general decline in the other perennials as well. Although the evidence is not so consistent as one might wish, it seems safe to say that the general pattern has been one of increase, and to attribute the two exceptions to peculiar local conditions, either in soil or in the cultural influences to which the localities have been subjected.

In a habitat that is marginal to the desert, where differences in slope exposure alternately create conditions favorable to Sonoran Desert and desert grassland vegetation, blue paloverde and mesquite have increased. Counts made separately of plants in three height classes indicate that the populations are sharply skewed in favor of young individuals in the case of both species (Plate 73; see caption to Plate 74).

Mesquite as an invader is most conspicuous at elevations above the desert where it has appeared during the past eighty years in the desert grassland and the oak woodland alike, on flood plains, in ravines, and on slopes. Within these zones, there are upland sites (Plates 22, 25, 27, 28, and 55) and lowland sites (Plate 54) at which mesquite was already growing sparsely or even densely (Plates 41 and 60) in the 1880's and 1890's. More characteristically, however, its presence today is largely or wholly the result of post-1880 establishment.

Invasion by mesquite is commonly thought to have begun just prior to the turn of the present century, and there are several published records which substantiate this belief (Griffiths 1903, 1910; Thornber 1910). The proliferation continues at mid-century, however (Glendening 1952; Plate 26), and there is evidence of incipient invasion in areas near Sonoita and elsewhere in southeastern Arizona (Plate 8). It is likely that mesquite will continue to gain in biomass at elevations lower than 5500 ft., its rough upper limit (Plate 1).

The legume family includes several genera other than *Prosopis* which are important as invaders within southwestern North America. Of these, *Acacia* is best represented in the photographs. *Acacia constricta* (white thorn) has increased conspicuously in the upper Santa Cruz Valley on lands once covered with grass (Plate 40). The closely related *Acacia vernicosa*, a Chihuahuan Desert plant, has invaded so extensively in the upper San Pedro Valley that the area is probably best considered Chihuahuan Desert today, although judging from the series of early photos of the valley (Plates 42–60), grasslands dominated the region less than a century ago.[1]

Acacia greggii (catclaw), the third species of the genus to have made gains in eighty years, is a common inhabitant of runnel communities in the Sonoran Desert (Plates 72, 73) but may occur in thin and scattered stands away from channels. This species has increased at several grassland sites in the vicinity of the Santa Rita Mountains (Plates 27–41).

The genera of the legume family that are active in vegetation dynamics in this area are similarly involved elsewhere. Shantz and Turner (1958) report the increase of several species of *Acacia* in Africa during recent years. Morello and Saravia (1959) list species of *Acacia* and *Prosopis* as woody invaders in arid parts of Argentina.

The list of woody invaders by no means includes *only* members of this family. Ocotillo, a nonleguminous plant characteristic of both the Sonoran and the Chihuahuan Deserts, has increased conspicuously at stations above the desert. In view of the species' establishment potential, it is perhaps remarkable that more plants are *not* found. Germinating seeds of ocotillo occur in prodigious numbers following the first summer rains. Mortality among the seedlings is typically high, however, with only one in every 10,000 to 100,000 surviving the first year (Shreve 1917a). Like many other woody perennials of the Sonoran Desert, the plant has a great potential to increase in density if conditions that favor the survival of its seedlings are improved even slightly.

Cottonwood, another non-legume, figures prominently in seventeen photographic pairs; of these, two-thirds show the tree to have increased, the remainder record either a static condition, a possible increase, or, in one instance, a decrease (Plate 41). The proliferation is remarkable, particularly in view of the paucity of recorded cases where the plant has decreased. Cottonwood establishment appears to be confined to the vicinity of streams with constant or intermittent flow, although old individuals may persist many years after all surface flow has ceased. Its increase in recent times is probably related to arroyo cutting. Sudden erosion first of all destroys the old vegetation, leaving open spaces temporarily free from competition. Secondly, it creates a variety of stream-side habitats where formerly the environment may have been more homogeneous. Finally, it eliminates marshy conditions by lowering the water table. Plate 7 shows young cottonwoods already occupying the area about a small spring after recent channel cutting had presumably effected a drop in the water table. The trees appear to have persisted to the present. Plate 57 shows a marshy open plain that today is channeled and heavily overgrown by such woody species as cottonwood and mesquite.

More than thirty species in all have taken part in the changes recorded by the plates. The ones discussed above have been singled out because of the frequency with which they appear. Changes in the others cannot be properly evaluated without more information, although undoubtedly the distributional dynamics of some are fully as significant as the ones discussed.

The extent to which the vegetation of the desert region has changed in the past eighty years is, as the plates attest, almost startling. Changes of this order are not at all to be expected as part of the normal course of events, and it now becomes necessary to examine the question of cause.

VIII

SOME HYPOTHESES OF VEGETATION CHANGE

Many studies have treated one or another of the aspects of vegetation change in the desert region, and reviews of the principal literature may be found in works by Humphrey (1958), and Parker and Martin (1952). The responsibility for the changes has been hotly debated. Determining cause is difficult because the effects of one force may be much like those of another, and several forces may have reached a critical intensity at about the same time. Thus, the decline of the grasslands—a change with which a lot of the past work has dealt—could stem from overuse by cattle, or from a climatic shift toward greater aridity, or from these two forces acting together. To complicate matters further, most studies have examined no more than the changes in a few species, treating them in isolation, set apart from related changes in other species that might serve to shed additional light on the problem. For both of these reasons the work that is based on experimental evidence often presents conflicting conclusions; the part that is purely theoretical contains so many contradictions that it defies any attempt at synthesis. Whether experimental or theoretical, the earlier studies have tended to concentrate on the parts played by four alleged agents, cattle, rodents, fire, and climate.

The effect of cattle. The bulk of the literature deals with the role of livestock in promoting shrub invasion, and emphasizes four ways in which domestic grazing animals have supposedly contributed to the spread of woody plants.

In the first place they have acted as disseminators, either by scattering viable seeds in droppings, or (as with the chollas) transporting the vegetative organs of plants from one place to another. As many as 1617 undigested mesquite seeds have been found in a single cow "chip" (Glendening and Paulsen 1955). Furthermore, many of these seeds are able to germinate, a trait which is closely related to the nature of their seedcoats.

The seeds of woody invaders of the desert region like mesquite, catclaw, white thorn, and *Acacia vernicosa* have hard, impervious coats and must be scarified before dormancy is broken and they can germinate. According to Morello and Saravia (1959) and Reynolds and Glendening (1949) the passage of these seeds along the alimentary tracts of domestic livestock may provide the necessary scarification.

However, Glendening and Paulsen (1955) also present data suggesting that before the full potential for germination can be reached, these seeds may require scarification outside the intestinal tract. Bacterial and fungal action may provide the additional increment, or it may be forthcoming from the mechanical abrasion received when the seeds are entrained in floodwaters together with gravel, sand, and rock.

Went (1957: 252) has suggested that leguminous trees occur so frequently along watercourses because of this factor. Glendening and Paulsen (1955) and Morello and Saravia (1959) have stressed the importance of floodwater scarification to mesquite.

So far as mesquite is concerned, then, cattle provide an effective means for dissemination. Moreover, they may accelerate the process of scarification. In both of these ways an increased livestock population may aid the spread of mesquite, and probably of related species as well.

But what of the next step in the plant's life cycle, seedling establishment? Here cattle evidently have a depressing effect, both on mesquite (Glendening 1952, Paulsen 1950) and on other woody plants (Shantz 1905). Presumably the animals consume by browsing the very seedlings whose introduction they have fostered in other ways. Since livestock often graze warm slopes earlier in the year and more intensively than less sunny, adjacent slopes (Humphrey 1962: 99), this factor may help explain the absence of mesquite on south-facing slopes in grasslands where it has invaded nearby slopes with a northerly aspect (Plate 34).

In summary, there seems to be little doubt that seeds of mesquite and other shrubs may be spread by livestock, and the increase in these species in recent years is *de facto* evidence that many of the seedlings survive and mature in spite of browsing pressures. What of the timing of the encroachment by woody plants?

In most places in southeastern Arizona it coincides with the period of heavy grazing that began in the 1880's. A great deal has been made of this coincidence, but, as has already been pointed out, none of the studies look farther back than the 1880's or farther south than the present International Boundary, and few of them even recognize that the region had a history prior to the Mexican War. A substantial amount of grazing in southeastern Arizona in the 1820's and 1830's evidently had no ill effects; nor apparently did the heavy grazing in Sonora from 1700 to 1880.

A second way in which cattle have allegedly contributed to shrub invasion is by "opening up" grass communities that were able in the past, because of their "closed" nature, to repel invasion. A tinge of anthropomorphism attaches to the notion of competition among plants, and the ecologist is never wholly comfortable in advancing an argument like this, since he knows so little about the specific mechanisms that comprise competition, if such a thing can be properly said to exist in the plant world. But grazing, it has frequently been suggested, performs this "opening-up" function, and is the primary cause of woody intruders in the grassland (Whitfield and Beutner 1938, Whitfield and Anderson 1938).

Several specific mechanisms have been mentioned that might accomplish this "opening-up": the grass mat may be so impoverished by grazing that seeds from shrubby species have readier contact with the soil; the amount of shade may be reduced, allowing shrubs to become established that formerly were excluded because they require full sunlight for seed germination or seedling establishment; the competition for soil moisture may be lessened, allowing woody seedlings a better chance to reach maturity.

Glendening and Paulsen (1955) examined the effects of herbage density on seedling establishment by planting mesquites in two series of plots, one artificially cleared of its dense grass cover, and the other with its herbage intact. The grass cover, they found, sharply reduced seedling establishment. Although the soil moisture content of the surface layer (0–6 in.) was generally greater on the partially shaded, grass-covered plots, at greater depths (6–18 in.) it was greater on the bare plots, thus favoring plants like mesquite with deep-rooting habits. The pattern of seedling losses suggested that drought, acting indirectly through imperfectly understood mechanisms, caused most of the mortality. Glendening and Paulsen emphasized the complex nature of the problem and concluded that, while a perennial grass cover limited the establishment of mesquite seedlings, the reason could not be determined from field studies.

Still a third way in which cattle may contribute to the spread of shrubs is by removing a large part of the grass cover, thus reducing the fire-carrying capacity of the range and tending to weaken fire's role in suppressing woody species. This will be discussed under the effect of fire suppression.

Fourthly and finally, grazing may tend to reduce the moisture content in the upper layers of the soil: by compacting the surface, thus reducing infiltration; by removing the grass cover and litter, thereby decreasing infiltration and increasing evaporation. The lesser soil moisture values, the reasoning goes, favor species like mesquite with a long tap root that can draw on deep-seated moisture, and tend to discriminate against the shallowly and intermediately rooted grasses. The work of Glendening and Paulsen (1955) provides the only experimental evidence for the desert region, and it does not directly implicate cattle as the agent responsible for reduced soil moisture.

The effect of rodents and jack rabbits. The place of rodents and other small animals in this discussion rests on two alleged relationships: one, between them and plants; the other, between them and man. The first, that they have an impact on plant life through their browsing and foraging, has been examined in some detail by a number of writers, and rests on an acceptable body of evidence. The second, that they have increased in number in recent years, either because white man has killed off some of their predators—snakes and coyotes—or because they find man's company congenial in other respects, has been tested scarcely at all and is generally assumed rather than demonstrated. And while it seems a reasonable assumption in some cases, it is by no means so in all. Between Tucson and Dos Cabezas Mountains, for example, a traveler in 1860 wrote that "thousands of prairie dogs were in sight of the road, and as we passed through their villages we could see hundreds of them wiggling their tails as they rushed into their holes" (*Daily Alta Californian*, July 4, 1860). These villages have vanished. Against the greater population of some species must be set the smaller population of others. It is far from clear what the net effect of the white man has been, and only with this important qualification can the first relationship, that of rodents and rabbits to vegetation, be examined.

The studies devoted to these small animals show in general that they are important foragers and may, on deteriorated range land, exercise a potent ecological force. However, they are probably not responsible for initiating the deterioration of a range. Once that process has been set in motion by other factors, and a grassland is already on the way to becoming depauperate, certain of the species may then increase in number, further contributing to the spiral of decline by promoting the establishment of woody plants like mesquite. Paradoxically, at this very time the plant life is subject to greater browsing pressure.

In one study in southern Arizona, Taylor (1936)

reported that the antelope jack rabbit, the banner-tailed kangaroo rat and several species of lesser importance removed about 30 per cent of the total vegetation. Impressive as the figure is, it may merely indicate that the animals are most abundant where the vegetative cover is already sparse.

Taylor *et al.* (1935) found the jack rabbit population to be largest on grazed lands. Similarly Reynolds and Glendening (1949) found an inverse relationship between the number of Merriam kangaroo rats per acre and the density of perennial grasses. The same relationship held even where rats had access to adjacent areas with a dense grass cover, protected from cattle and rabbits. The authors speculated that a preferred food, grasses with large grains, may have been present in greater abundance on the sites with poor cover. Or the rats may have liked the lightly covered areas because the open terrain gave them greater freedom of movement.

In a study of black grama grassland in New Mexico, Norris (1950) found rodent control to be unnecessary on well managed ranges; on denuded ranges dominated by mesquite and snakeweed the perennial grass cover improved only with protection from the small herbivores. Taylor *et al.* (1935) concluded that the primary responsibility for depleted ranges lay with livestock, the rabbits and rodents playing a secondary role.

In the establishment of woody species the small mammals evidently give rise to two opposing forces. On the one hand such species as the Merriam kangaroo rat may aid mesquite establishment, first by scarifying the seeds by gnawing them to test for soundness, then storing the unused ones in caches where they are able to germinate (Reynolds 1950). Seedlings from such sources are most numerous on poorly grassed sites, where the rat population is highest (Reynolds and Glendening 1949).

On the other hand, the increased browsing that arises with an increased population of the animals may be deleterious to shrubs. Over a period of fifteen months Paulsen (1950) found a 47 per cent mortality rate among mesquite seedlings on plots from which cattle and rodents were excluded; of 95 per cent on plots where rodents, or rodents and cattle, were allowed free access. After analyzing stomach contents, in a study conducted on the Santa Rita Range Reserve south of Tucson, Vorhies and Taylor (1933) estimated that 36 per cent of the annual diet of the antelope jack rabbit consisted of mesquite; and of the California jack rabbit, 56 per cent. In spite of these high values they found no evidence on the Range Reserve that browsing handicapped plant growth. In the upper San Pedro Valley and eastward into New Mexico, however, injury was severe enough to impair growth.

It is virtually impossible to assess the net effect of the many forces acting here. Rodents and rabbits may contribute to the increase of certain woody species which, in turn, they then destroy. If man has, in fact, abetted an increase in rats, mice, and rabbits by controlling snakes, coyotes, and other predators, what ought to be the effect on the vegetation? It is difficult to say. If these small mammals reach great numbers only on depleted ranges, basically the question is what caused the depleted ranges. And this is the problem examined by this book.

The effect of fire suppression. Some aspects of the fire hypothesis have already been discussed. Briefly it states that, acting as conservationists and in the interest of protecting their property, Anglo-Americans have sought to control fire. In recent years their efforts, coupled with a weakening of the grass cover through overgrazing, have appreciably lowered the incidence of burning. Whereas fires used to occur frequently, periodically eliminating woody plants from the landscape, in recent years such seedlings have become established in large numbers.

In order to be effective in stemming the encroachment of woody species, fires must be frequent enough and hot enough to eliminate all woody seedlings, many of which have strong propensity to sprout from the root crown when the shoots are killed. In the desert region, then, fire as a factor in vegetation dynamics should be most important in the oak woodland and desert grassland, where grasses provide abundant fuel, rather than in the desert proper, where fuel is insufficient to carry a fire (Humphrey 1962).

The fire hypothesis has been hotly argued. Much of the work dealing with it has been of a conjectural nature, involving subjective judgments. Moreover, there has been some confusion of the two basic questions involved.

The first of these, one for the ecologist and the range manager, has to do with fire's effectiveness in controlling shrubby plants. Does recurrent burning suppress them? The second, falling more within the province of the historian and anthropologist, concerns the historical frequency of burning. Before Anglo-Americans arrived, how frequently did grassland fires occur in the desert grassland? The tendency has been for a researcher to answer one of the questions to his satisfaction, and to assume that he has answered both. This is clearly not the case. Fires may be effective in controlling shrubs, and may have occurred frequently in the past. Fires may be effective in controlling shrubs, but may not have occurred frequently in the past. Fires may not be effective in controlling shrubs, but may have occurred frequently in the past. Fires may not be effective in controlling shrubs, and may not have occurred frequently in the past. The two questions are independent of each other, and separate answers must be provided to them. Depending upon the answers, one of the four combinations listed will prevail. The first establishes the fire hypothesis. The second denies the fire hypothesis, but grants that fire may have been locally important to the vegetation when and where it did occur. In the third and fourth cases, the fire hypothesis is inadmissible.

To the first question there are better answers than to the second. Several workers in the desert region have recently obtained quantitative data that are useful in judging fire's effect in maintaining shrub-free grasslands. Humphrey (1949) has shown that June fires, immediately before the summer rains, virtually eliminated burroweed, a small, worthless intruder on deteriorated ranges. He presented no quantitative data for mesquite, but many of these plants were also killed.

Reynolds and Bohning (1956) have described the results of burning a desert grass-shrub range in southern Arizona: fire killed up to 88 per cent of the burroweed, but no more than 9 per cent of the mesquites. All of the mesquites killed were small, having stems less than 6 in. in diameter. The fire burned through the area erratically, a factor which may explain the low rate of mortality among the latter species. Black grama, one of the most important grass species in the desert grassland, was seriously damaged by the burn, and had not recovered by the end of the study. Humphrey (1949: Table 2) found similar but less conclusive evidence of a harmful effect on black grama and other grass species.

Glendening and Paulsen (1955) have reported the effect of experimentally burning mesquite where baled hay was spread over the plots to supplement the rather thin, natural fuel. They found some indication that fire injury varied with the season, there being a higher kill during June than in the fall or winter months. Mortality varied inversely with plant size, the greatest injury occurring in the small size classes. The highest mortality rate, 60 per cent, was found among plants with stem diameters smaller than 0.5 in. As the result of these studies, they concede "that fires may have been of some importance in holding mesquite in check prior to settlement of the Southwest" (1955: 48), a conclusion which does not follow at all from their experiment, and which illustrates the frequent confusion of the two questions even among careful workers.

To the second question, whether fire used to occur frequently in the desert grasslands of the region before the advent of Anglo-Americans, no satisfactory answer has been given. Humphrey (1958), who has investigated the historical aspects of the matter more thoroughly than anyone else, concludes that burning used to occur frequently. However, his evidence consists either of statements by secondary authorities

—a form of evidence inadmissible in historical research—or statements from primary sources that document fires elsewhere, in Texas or the Midwest. These latter are similarly inadmissible. If burning occurred frequently in the grassland of the desert region the abundant historical materials pertaining to the area should show it without recourse to the irrelevant question of fires elsewhere.

Branscomb (1958) states that in his review of historical materials "no reference to fires in the Jornada area [of New Mexico] were found." However, he adds, "it has been established that range fires occurred periodically in the desert grassland during the early years of, and before white settlement (Nuñez 1905, Humphrey 1953)," and the Jornada area "seems to have been an extensive grassland" at one time. Therefore, the implication is, range fires occurred periodically in spite of his having found no mention of them. Of the authorities he cites who have "established" this fact about the desert grassland, Humphrey is, of course, a secondary—or perhaps, more correctly, a third-hand—source, and it is not clear why Alvar Núñez Cabeza de Vaca is even mentioned. His route across the American Southwest is still being argued by historians. If one follows Hallenbeck's view (1940), Cabeza de Vaca did indeed traverse the desert grassland; however, the only mention his journal makes of grass burning pertains to an area in central Texas.

So far as is known, the six quotations in Chapter III are the only citations anywhere in the literature, from primary sources, that mention fires in the desert region. As has been pointed out, two of the six involve fires in the desert valleys, not in the grasslands; three of the six mention grassland fires set by the Anglo-American travelers themselves. One deals with a fire seen in the distance. From a study of twenty-two travel journals one can only conclude that emigrants traversed burned areas in the grassland very seldom.

For now the best answer to the two questions is unsatisfactory, but is that (1) fires may be effective in controlling shrubs, and (2) they did not occur frequently in the desert grassland in the past. This combination of answers denies the fire hypothesis, but grants that fires may have been locally important to the vegetation of the zone when and where they did occur. It should be emphasized that this conclusion applies only to the grassland of the desert region. Because of differences in the flora, in the climate, in the terrain, and in the habits of the native peoples from one region to another, each part of the world has to be viewed as an individual case.

Climate. Climatic change has already been touched on in connection with arroyo cutting. A number of students have suggested its applicability to the problem of plant invasion as well. There is general agreement that increased aridity tends to favor deep-rooted plants like some of the shrubs, and to depress the grasses, which combine more mesic requirements with a relatively shallow root system.

Insofar as it is synonomous with drier soil conditions, however, aridity can be produced by factors other than climate and does not necessarily mean the same thing as deficient rainfall. Anything that tends to reduce the amount of water infiltrating the earth—diminished ground cover, for example, or mechanical compacting—may have the same effect as decreased rainfall. Several such factors are associated, directly or indirectly, with man's activities, and a number of writers have postulated the phenomenon of "desertification"—increased aridity as the result of human habitation in an area.[1] It has apparently occurred in arid and semiarid areas throughout the world. In South Africa it goes by the picturesque name "the creep of the Karoo." Almost invariably it is sooner or later mistaken for a sign of changing climate.

The problem with respect to the desert region is twofold: first, to distinguish desertification from climatic change; secondly, assuming that there is a residuum of increasing aridity that cultural factors do not account for, to isolate in the weather records the climatic variable responsible. This is by no means as simple as might be supposed.

The use of old weather records. In the first place, the weather records for the region before 1900 are inadequate. Where they do exist (Tucson has the longest record, going back to 1867), they are incomplete or unreliable. Missing data are legion. Neither instruments nor shelters used to be standardized; different instruments may have been used for different portions of the record. The stations frequently were—and still are—moved from location to location without a change in the station name.

In the second place, it is difficult to deal with the problem of trend. Given a time series of average annual rainfall amounts progressing steadily upward or downward, there is no difficulty. But in the case of the desert region with its high temporal variability, one is apt to have instead a series in which random year-to-year variations obscure any apparent directional tendency.

Smoothing is frequently resorted to as a way of surmounting the difficulty. But it has its perils.[2] Various statistical tests exist to determine whether trend is present, but unless a systematic tendency is quite pronounced, it will not override the noise in time series as short as those available for the desert region. Thus, while "statistical significance" is a valid and important probability statement of the degree of departure from randomness, it may have little or no relation to what constitutes "vegetative significance." Indeed, "trend" itself need not be related to plant welfare, if one defines the term narrowly enough. It is possible to describe a time series of annual rainfall amounts free from any statistical tendency at all toward average increase or decrease that, if it represented an actual sequence of events, would have brought the vegetation to near extinction.

Trends in rainfall. Is there any reliable indication that the climate of the desert region has, in fact, changed during the past century?

In the early records for New Mexico Leopold (1951a: 351–52) has noted a shift toward fewer small rains and more large ones.

Such a circumstance must have been conducive to a weak vegetal cover and relatively great incidence of erosion. That the modern epicycle of erosion began in the Southwest about 1885 is well established. . . . We see, then, that not only was grazing tending to promote erosion at that time, but meteorological conditions were more conducive to erosion than during the period of the present generation. Thus there is established concrete evidence of a climatic factor operating at the time of initiation of southwestern erosion which no doubt helped to promote the initiation of that erosion.

An analysis of the rainfall records since 1895 for five stations in southeastern Arizona shows some affinities to the curves presented by Leopold. Whether there is general agreement for the period before 1895 is a question on which judgment must be deferred. What effect such a shift would have on the vegetation is a most difficult question and can be answered at present only in an unsatisfactory qualitative manner.

From tree-ring studies, Schulman (1956: 67) has inferred the existence of a severe drought since about 1921 in the southern Gila River Basin, where the San Pedro and Santa Cruz Rivers are located:

It appears highly likely, in view of the general parallelism with the chronologies in Colorado and Utah, that this is the most severe drought since the late 1200's. If this is correct we have, then, a direct climatic explanation for the high mortality which has been observed in recent years in *Carnegie* [sic] *gigantea* (sahuaro) and in the low-level, woodland pines *Pinus leiophylla* var. *chihuahuana* and *P. engelmanni* (*P. apacheca*).

In a definitive analysis of precipitation between 1898 and 1959 at eighteen stations in Arizona and western New Mexico, Sellers (1960: 85) has reached a similar conclusion—that there has been a downward trend in rainfall averaging about one inch every thirty years. "The decrease has occurred almost entirely during the winter months, although there has also been a slight decrease in summer. None of the trends is statistically significant." Since 1921–24, the 20-year average annual precipitation in Arizona and western New Mexico has decreased by about 25 per cent (*ibid.,* p. 85).

Von Eschen (1958) has noted comparable events for several New Mexico stations; Hubbs (1957), treating the problem in a general way and supporting his conclusions with a variety of evidence, has postulated a long-term trend toward aridity throughout the desert region.

Trends in temperature. During the latter half of the nineteenth century it became apparent in the higher latitudes of the Northern Hemisphere that glaciers were retreating, that the timberline was advancing upward and northward, and that the waters of the North Atlantic were becoming warmer (Ahlmann 1949). Since 1920 an increasing body of literature has dealt with phases of the "recent climatic fluctuation," as it has come to be called; [3] the extent of the warming is now known to be worldwide. On the basis of climatological data, Willett (1950: 205–6) concluded that the increase varies somewhat with latitude, but in general, dates from 1885 and amounts to about 2.2° F. in winter mean temperatures, 1.0° F. in annual mean temperatures.

After updating Willett's study and weighting differences areally to secure a better balance among latitudinal belts, Mitchell (1963: 161–62) reported "rather uniform [*i.e.* uniform temporally, not geographically] rates of warming from the 1880s to the early 1940s and a marked tendency for cooling since the early 1940s." His results agree well with Willett's, although his average world-wide increases are of smaller magnitude, about 1.6° F. for winter and 0.9° for annual mean temperature. Callendar's results agree substantially, except for the southern hemisphere (Callendar 1961, Landsberg and Mitchell 1961). Butzer (1957) has associated a tendency toward aridity with the temperature changes in North Africa. The fluctuation in the United States has been studied by several workers, including Kincer (1933 and 1946) and Page (1937).

In view of the extensive documentation for other parts of the world, it would be surprising if weather records for the desert region did *not* show analogous trends. Analysis shows that the region has indeed participated in the general warming, and that the local temperature curves bear a close relationship to those for other parts of the world.

Figure 13*a* shows the mean annual values at Phoenix-Tucson-Yuma for the period between 1898 and 1959, together with three curves redrawn from Kincer's work (1946). At the three Arizona stations there was a mean increase of about 2° F. in mean annual temperatures between 1911–20 and 1950–59. Since part of the rise may have been associated with the urbanization of the Tucson and Phoenix areas,[4] a more elaborate time series has been constructed based on the records from 1893 to 1959 for eighteen Arizona stations located only in rural areas and small towns; this appears as Figure 13*b*.[5]

For the six hot months (May–October), there was an increase of 2.4° F. in the period between the decades ending in 1913 and 1940. For the six cold months, the increase amounted to 1.75° F. for the period from the decade ending with the winter of 1912–13 to that ending with the winter 1942–43.

Because the two Arizona curves are similar to Kincer's for St. Louis and Minneapolis–St. Paul, dating back to 1846, it seems likely that the upward trend in the Southwest actually began earlier than the reliable period of record, since 1893. Mean annual temperatures in Arizona may have risen by 3 or 3.5° F. since the 1870's.

Could the amount of warming actually observed account for the vegetative changes that have taken place? Initially the increase seems rather insignificant, and without a precise knowledge of the temperature responses of the plants that are involved, it is difficult to give a definite answer. However, some general indication may be gained from the spatial relationship that prevails between temperature and elevation in southern Arizona. The nineteen stations listed in Figure 4 show an average rise in mean annual temperature of about 1° F. for every 300 ft. decrease in elevation.

A temporal increase of 3° F. in mean annual temperature, then, would be of the same order as the spatial change occasioned by dropping about 1000 ft. in elevation. Since one can go from the oak woodland across the desert grassland to the upper edge of the desert in this span, it appears that the observed temperature increase since 1893 might indeed have significance for the vegetation.

On the basis of Sellers' careful work (1960), it is possible to say, moreover, that the period since the turn of the century in Arizona has been characterized by a slight decrease in summer precipitation and by a marked decrease in winter precipitation. These trends in rainfall tend to reinforce many of the effects of higher temperature.

In Figure 14 Sellers' data, smoothed by a ten-year running mean to conform to the practice with the other curves,[6] are grouped, by season, with the average temperature data for eighteen Arizona stations. The latter have been truncated by five years to make the time intervals correspond. During the six cold months (November–April), temperatures have risen, and rainfall has fallen sharply. During the six hot months (May–October), temperatures have risen and rainfall has fallen slightly.

On the basis of these configurations one can reasonably postulate that since 1898 (1) there have been trends toward greater evaporation and less soil moisture, (2) there has been a tendency for the life zones of the desert region to retreat upslope.

From the climatological data available, these tendencies can only be extended back to 1898. On the basis of the similarity of the temperature trends in the

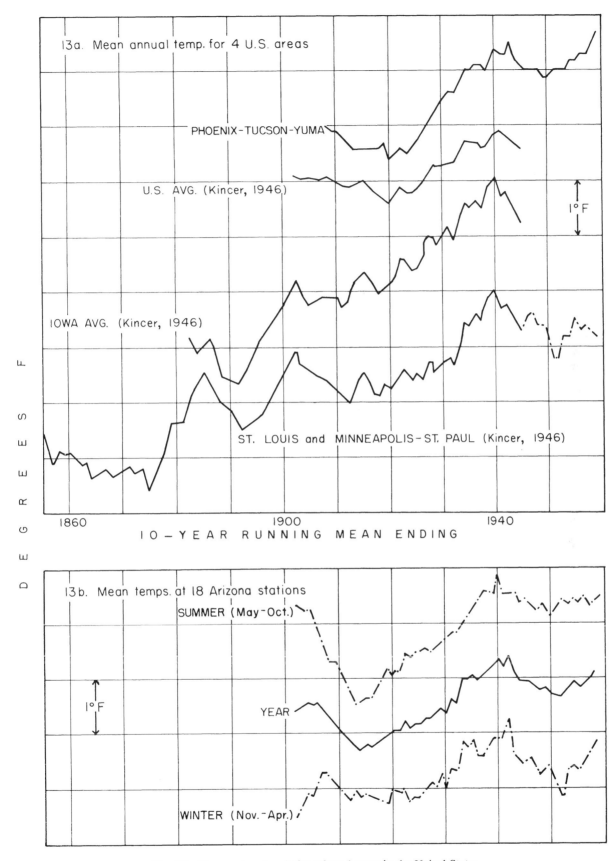

FIG. 13—Temperature trends for selected areas in the United States.

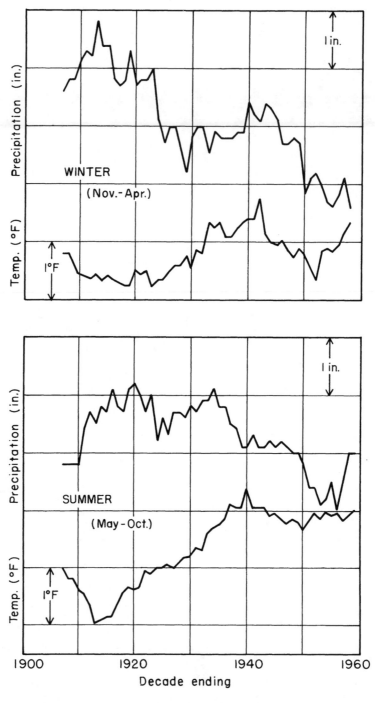

FIG. 14—Ten-year running means comparing mean seasonal precipitation at 18 Arizona and western New Mexico stations (after Sellers 1960) with mean seasonal temperatures at 18 Arizona stations.

[282]

desert region to those in other parts of the country having a longer period of reliable record, however, one can reasonably infer that the warming may date from the 1870's.

To the other hypotheses discussed, pertaining to cattle, rodents, fire suppression, and other aspects of climate, must be added this one: that the higher temperature and lower rainfall since 1898 have made their contribution to the changing vegetation of the desert region.[7] These hypotheses will now be evaluated in light of the evidence that has been accumulated in the earlier chapters.

IX

CHANGE AND CAUSE

The first chapter of this study surveyed the principal climatic and ecological features of the desert region; the second and third reviewed the region's history, emphasizing the cultural factors that bear on the problems of vegetation change and arroyo cutting. In the following three chapters photographic pairs spanning periods up to eighty years in length were presented and analyzed to determine the degree and nature of the vegetation changes that have occurred in the region since 1880, and to ascertain which plant species were involved. Chapter VII summarized some of the patterns vegetation change has followed. In Chapter VIII the more prominent hypotheses that attempt to account for the changes were reviewed. With this background it now becomes possible to evaluate the hypotheses and to draw some tentative conclusions about causation.

The effect of cattle. The principal historical and photographic evidence bearing on the question of cattle as the agent of change is equivocal: there is evidence that overgrazing acted as a primary cause; there is evidence that it did not. On both sides of the question, the timing of pairs of events is important. What is not clear in any single case is the degree to which coincidence is involved; a cause and effect relationship need not follow from a temporal correlation.

There is a close association in time between the rapid expansion of the cattle industry in southeastern Arizona in the 1880's and the onset of arroyo cutting and vegetation change in that area. The association invites the conclusion that one event is causally related to the other; however, without additional evidence such a conclusion is unwarranted.

To some extent, scientific studies reinforce the notion of a link by indicating the mechanisms that might be involved: that livestock scatter viable seeds of some shrubby species in their droppings has been established by the work of Glendening and Paulsen (1955). That they contribute to scarifying the seeds of some shrubby species has been shown by Reynolds and Glendening (1949) and Morello and Saravia (1959). Plate 8 shows what is apparently an example of this joint effect near old Fort Crittenden, where a marked mesquite invasion has taken place at the site of an old corral. Mesquite establishment

probably postdates heavy use of the corral, since heavy trampling and browsing would preclude establishment.

That cattle graze is self-evident, and that a severe stripping of the grass cover contributes in a poorly understood way to the establishment of shrubs seems clear from the work of Glendening and Paulsen (1955). That removal of the plant cover reduces infiltration and therefore tends to result in lower soil moisture values also seems indicated. That such a stripping occurred is borne out by the photographic evidence: Plates 22 and 43 show ranges in southeastern Arizona as they appeared in 1890 and 1891 respectively. It is clear from them, and from the abundant historical evidence as well, that the ranges were virtually denuded of their cover in the late eighties and early nineties. What is not clear is the extent to which overgrazing alone was responsible, and the extent to which drought. In contrast to forests, where the longer lived individuals dampen the community's response to minor fluctuations of climate, grasslands vary greatly in composition from season to season and from year to year (Coupland 1959). A relatively fast response might be expected to unfavorable conditions generated either by grazing or climate. It seems likely, at least in the case of Plate 43, that both factors combined to produce the conditions shown.

Against this evidence for the cattle hypothesis must be set several arguments against it. Cattle are browsing animals as well as grazing animals. The work of Glendening (1952), Paulsen (1950), and Shantz (1905) indicates that cattle consume the seedlings of some shrubs, thus depressing establishment rates. This qualitative argument is offset to some extent by the simple quantitative observation that shrubs have managed to increase in recent years despite a large livestock population.

More serious are the historical objections. Large-scale cattle-raising in Sonora commenced nearly two centuries before 1880. There is no evidence that significant vegetation changes accompanied them. There is good evidence that a seemingly related phenomenon, arroyo cutting, began in northern Sonora at the same time as in southeastern Arizona, about 1890. This argument, involving the *lack* of temporal correlation between cattle-raising on the one hand and vegetation change and arroyo cutting on the other, undermines much of the significance attached to their correlation in southeastern Arizona in the 1880's. However, it is not conclusive, because no figures are available to document the size of the livestock population in Sonora at the pertinent times.

Two similar arguments may be advanced, but are open to the same qualification, that precise quantitative data are lacking:

A transient but substantial development of ranching in southeastern and south-central Arizona in the 1820's and 1830's evidently had no effect on the vegetation or hydrology of the area.

Around Tumacacori Mission (Plate 40) continuous grazing has taken place since 1700, probably reaching a peak in the early nineteenth century. The transition from grassland to desert-scrub vegetation occurred at some time after 1891.

Moreover, there are certain examples of change that cannot be explained adequately by the cattle hypothesis. One of these concerns the oaks at Gird Dam (Plate 49). The general retreat of the oak woodland upslope is consistent with the contention that grazing tends to produce "desertification," because it tends to lower infiltration. But this effect would be felt primarily in gentle terrain with a great deal of grass and with few rocks. In the case of the Gird Dam picture, the site is a rocky, boulder-laden ridge. Cattle tend to avoid such areas; a large part of the surface is stone, not soil: its infiltration characteristics could be changed only slightly even if it were heavily grazed and trampled. Yet the oak mortality is high.

Another example involves MacDougal Pass (Plates 82–83), an arid, hot location toward the lower edge of the desert, where there has been a striking mortality among blue paloverde, *Lycium,* mesquite, creosote bush, and big galleta. Here the grass cover has probably never been dense. The soil is sandy, absorbs water rapidly, and compacts only slightly. Cattle, although they have grazed the area, probably have had little effect on the infiltration characteristics of the soil.

The most decisive case concerns MacDougal Crater (Plates 85–88) and the Melisas Islands (Plates 96–97). Almost certainly neither of these sites has even been visited by domestic livestock. Yet at both there have been significant changes. The crater presents a striking panorama of dead and dying vegetation: creosote bush, foothill paloverde, mesquite, and big galleta. On both of the small islands in Guaymas Harbor there has been a significant increase in the number of *cardones,* the giant cactus of northwestern Mexico.

The net inference to be drawn from these conflicting and not really conclusive bits of evidence for and against the cattle hypothesis inevitably comes down to a matter of individual judgment. Our conclusion—if, indeed, that strong a word is warranted—is that livestock have made important contributions to the changing vegetation of the desert region, but have not been the primary agent of change. The historical evidence in connection with Sonora from 1700 to 1880 and southeastern Arizona during the 1820's and 1830's seems to indicate that large-scale grazing need not be accompanied by vegetative or hydrologic change. The photographic evidence from MacDougal Crater and the Melisas Islands indicates that important changes have taken place at protected sites where livestock have never been. It seems likely that cattle have had more to do with the changes in the desert grassland than in the other zones; there their influence may have been considerable.

The effect of rodents and jack rabbits. There is no photographic evidence relevant to the rodent—jack rabbit hypothesis, and only the general historical fact that Anglo-American settlement immediately preceded the onset of arroyo cutting and vegetation change. There is no conclusive evidence that these small animals have increased in number, although this is widely alleged.

To the extent that settlement resulted in the suppression of predators; the suppression of predators, in an increased rodent and jack rabbit population; their increased population, in a greater impact on plant life, the hypothesis may be valid. The chain of causation is tenuous, and depends more on ingenious reasoning than observation. The experiments on which the hypothesis is based establish merely that rodents act as carriers and planters of seeds and vegetative joints, or that they and jack rabbits browse some plants. It is not clear what the net effect would be on a species like mesquite that is both disseminated and browsed. It is not apparent that the net effect should be a decline in some plant species and an increase in others. It is not clear why, in isolated areas like the Pinacate Mountains, where predator suppression has surely been less than around highly developed urban areas, vegetation change is nevertheless of a comparable magnitude.

The work of Reynolds and Glendening (1949) and Taylor *et al.* (1935) suggests that, in any event, the effect of rodents and jack rabbits is felt most strongly on deteriorated ranges. In this case their activities can be regarded as a secondary complication, not as a primary cause of vegetation change. The question should be left open, but there is no evidence that rodents played any significant part in initiating the ecological events following 1880.

The effect of fire. The fire hypothesis can best be approached by breaking it into two component questions, and examining each of them separately. The first, the effectiveness of fire in controlling shrubby plants, has been the subject of several investigations. Humphrey (1949), Reynolds and Bohning (1956), and Glendening and Paulsen (1955) have all demonstrated that burning kills many shrubby seedlings. There is some doubt that recurrent burning could produce a shrub-*free* grassland under the conditions prevailing in the desert region; certainly there is no reason to doubt that it can suppress establishment, keeping the rate lower than it would otherwise be.

The second question, whether fires used to occur frequently in the desert grasslands before the coming of Anglo-Americans, has not been answered so satisfactorily. The principal historical and photographic evidence relating to the question is as follows:

There is qualitative evidence that fire was used as a hunting tool by one group among the Apaches, but no quantitative evidence that either natural or man-induced burning occurred often enough or over wide enough areas to maintain the grassland in a shrub-free condition.

An examination of the frequency with which mid-nineteenth century accounts of travel across the desert grassland mention fire or burned areas lends no support to the notion that burning used to occur extensively. The evidence, in fact, supports the reverse of the fire suppression hypothesis—that man-made fires increased with the coming of Anglo-Americans.

Plates 46 and 47 document a severe invasion by mesquite in the Lewis Springs area. The earlier photographs, made in 1891, show the vegetation to be mesquite-free grassland studded by palmillas, highly inflammable plants in which burning, if not fatal, commonly induces the formation of multiple stems. Since many of the palmillas were at least fifty years old and had shaggy, single stems at the time of the old photographs, it is unlikely that fire had visited the area for many years. Without recurrent burning this particular locality was mesquite-free. There is no reason to suppose that the same factors that maintained an open grassland here could not have operated equally effectively in other parts of the region, or throughout the region.

The studies that have dealt with the relation of fire suppression to vegetation change in the desert region have without exception been concerned with either the desert grassland, or a life zone not discussed here at all, the pine forest. The changes in these zones have not been considered in relation to changes in the zone that lies between them, the oak woodland. The most conclusive argument against the fire hypothesis arises from precisely such a consideration.

The widespread occurrence of oak savannas before 1890 bears an anomalous relationship to the desert grassland if, as many workers believe, recurrent burning maintained the latter zone in a brush-free condition. One is forced to reckon with the problem of two life zones, intimately intermingled along their border, linked by a continuous grass matrix capable of conducting fire, one zone kept free of woody species through burning, yet the chief characteristic of the other being its woody dominants.

Some workers have asserted that the savanna itself may be a fire-induced form of vegetation,[1] and it seems easier, for the sake of advancing the argument, to grant this than to try to explain the circumstances by which recurrent burning would occur everywhere below the upper edge of the grassland and nowhere above.

The effect of fire on the oak species that are present has been described by several observers. There seems to be general agreement that young individuals are usually killed outright, although larger ones may escape with varying degrees of injury.[2] To produce an oak savanna, then, burning must occur frequently enough to keep the woodland thinned of brush and of *most* young oaks; yet not so frequently as to kill off *all* young oaks. The latter condition would ultimately render the area completely treeless, since there would be no replacements for the fire-resistant adults as, in the normal course of events, they died.

Looking now at the question of how the next zone below, the desert grassland, can be kept free of trees and shrubs by the action of fire, one finds that the effect of burning on mesquite is qualitatively similar to the effect on oak: the young trees are killed outright; mature ones may survive with more or less injury.[3]

To maintain a mesquite-free grassland, then, burning must occur frequently and widely. It must occur often enough to kill *all* of the mesquites while they are still young, since, once grown, they would survive fire. Furthermore, in their adult form, competing for soil moisture, they would gradually weaken the surrounding grass community, rendering it incapable of carrying fire, thus creating an expanding, sparsely covered area in which other mesquites could become established (Parker and Martin 1952: 33–38, Humphrey 1958: 52).

But how can these needs be reconciled with the needs of the woodland, adjacent to the grassland and joined to it by a common fire-conducting matrix of grasses? In the one case burning must occur frequently enough to keep all mesquites suppressed; in the other, infrequently enough to permit some oaks to become established. There is no historical reason to suppose that the requisite frequency of burning in the two zones followed such a pattern. In the first place, they were linked by a common carpet of grass that could conduct fire from one to the other. Furthermore, there is no *a priori* basis for supposing that Indian hunting activity—therefore, aboriginally induced burning—was any less extensive in the woodland than in the grassland. Thirdly, there is good reason to suppose that lightning strokes—therefore, lightning-caused fires—were more frequent in the higher zone than the lower. Finally, fire propagates upward easily and downward with difficulty. Flames originating in the grassland would spread with relative ease to the woodland; woodland fires, with relative difficulty to the grassland.

It is possible to postulate a weaker grass cover in the woodland, less capable of supporting combustion and therefore rendering the zone more susceptible than the grassland to the maintenance of woody species. This explanation fails, however, when one turns his attention away from considering the two life zones as static features of the old landscape and looks at the dynamic aspects they have exhibited in recent years. How, in terms of the fire suppression hypothesis, can one account for the recent invasion of the woodland by mesquite? It becomes necessary to postulate conditions in the past when fire, sweeping the one zone—the woodland—was at once able to eliminate all young mesquites, but to leave many young oaks for replacement purposes. There is no evidence that young oaks are more fire-resistant than young mesquites; even if they were, one has difficulty envisioning wild fires that everywhere were so nicely adjusted in temperature that they could perform this selective function over thousands of square miles without fail. Supposing the selection actually to have occurred, one is still left with the residual question of why, with the suppression of burning, the oaks have not proliferated as well as the mesquites.

If one starts with the initial assumption that the oak savanna, like the grassland, is fire-induced, there is no combination of circumstances that can explain the past existence of the two side by side in a brush-free condition. If, on the other hand, the savanna is not a fire-induced form, what used to keep it free of mesquite? In terms of the fire hypothesis it is possible to imagine an oak-mesquite woodland in the past, but not an oak woodland. Clearly factors other than fire suppression must be involved in the recent invasion of the oak zone by mesquite.

From this argument one can only conclude that fire was not the primary mechanism that used to keep the desert grassland and the oak woodland free from shrubs. Coupled with the more direct historical and photographic evidence, the chain of inference supplies a tentative answer to the second question: there is no reason to suppose that fires used to sweep the desert grassland frequently or on a large scale.

And on the basis of this answer one must reject the hypothesis that fire suppression has been a primary cause of the changes. At the same time one can readily grant the usefulness of fire as a tool in range management, and concede that fires, where they did occur in times past, were probably locally effective in keeping shrub establishment lower than it would otherwise have been.

The effect of climate. The evidence bearing on the hypothesis that climatic variation has brought about the changes of recent years is as follows:

The twenty-year period from 1875 to 1895 saw the inauguration of arroyo cutting in Arizona, New Mexico, Sonora, and Utah. White settlement in these states took place at various times from 1598 (in New Mexico) to the 1870's. Grazing commenced at equally diverse times. The heterogeneous history of the region makes it difficult to link settlement, and therefore cultural factors, with arroyo cutting; the uniform onset of erosion, on the other hand, points to the operation of a broad, regional factor like climate.

The historical facts accord well with the notion of climatic variation. The patterns of vegetation change revealed by the photographic pairs accord equally well. There have been striking trends toward a more arid vegetation, the dominant pattern being an upward displacement of plant ranges:

The lower edge of the oak woodland has been retreating upward in recent years. On the lower edge of their range where moisture is evidently the controlling factor, the two paloverdes and mesquite have retreated away from dry, hot habitats. On the upper edge of their range, where temperature is evidently the controlling factor, they have advanced. These movements are consistent with the hypothesis of drier, warmer conditions at all elevations.

Although the vegetative changes clearly point to drier conditions, aridification may be produced by man's activities as well as climate. Thus, while one hesitates to state categorically that climate is responsible for all of the aridification noted, in one instance, in MacDougal Crater (Plates 85–88), there have been important changes in a habitat undisturbed by man; and in isolated localities like Punto Cirio (Plates 92–94) and MacDougal Pass (Plates 82–83), where man's influence has at most been slight, there are also signs of greater aridity.

An association of two events in time, as has already been pointed out, does not necessarily indicate that they are related causally. And in this case, not even a loose temporal association between climatic variation and vegetation change has been shown. The possibility of doing so remains remote because good long-term climatological records for the desert region do not exist. Reliable curves for both temperature and precipitation can be safely constructed back to 1898 and these, smoothed by ten-year running means, appear in Figure 14. With fluctuations, winter rainfall has dropped markedly and winter temperatures have risen; summer rainfall has remained about the same, or has decreased slightly; summer temperatures have gone up sharply. In view of the relation that prevails between vegetation zones and the vertical distribution of temperature, the warming since 1898 has probably been large enough to affect the vegetation. Certainly the decrease in winter rainfall has been.

But there remains the problem of the earlier timing: to the extent that local warming and increased aridity can be associated with the general climatic fluctuation in the Northern Hemisphere, one can say that the trends date from about 1870. To what extent can the association be made? Temperature is a stable datum, and the Arizona curves look enough like those for the Midwest (Figure 13) that one can infer, with a reasonable expectation of being right, that warming in the desert region is related to the larger pattern, and dates from the 1870's. But rainfall is spatially unstable. Especially and emphatically this is so in the desert region which has two seasonal regimes and large spatial variability. One can safely infer nothing beyond what the data show. What about the timing of the vegetation changes?

On the basis of the paired photographs one cannot assign a date to the beginning of any of the recent changes. One can say only that during the period spanned by any two pictures a change did or did not take place. From the observations by early botanists we know that mesquite invasion had begun by 1903 (Griffiths 1904), and by 1910 had become obvious (Griffiths 1910).

To the extent that arroyo cutting accurately reflects changing vegetative conditions it is possible to be more precise. Arroyo cutting began along many of the streams of the desert region in August, 1890. One can infer, then, that by 1890 the vegetation had been altered enough to affect runoff, but it is an uncomfortable inference, resting as it does on the unproven assumption that a change in the vegetal cover inaugurated arroyo cutting.

The direct evidence is better, but still slight. Plate 20 shows young mesquites springing up in 1887 in the grassland on Proto Ridge; Plate 55 shows the same thing for mesquite in Walnut Gulch in 1890. The invasion by this species, then, may have begun somewhat earlier than 1890, possibly around 1880. The two events—a climatic shift and an alteration in the vegetation—can be loosely and generally associated in time, although the evidence is less than impressive.

In spite of the uncertainties connected with it, retention of the climatic hypothesis seems warranted. It conflicts with none of the known facts; on the positive side it seems capable of explaining more of the facts than any of the other hypotheses. The chief objections to it are that the outline of climatic variation remains vague in the critical period between 1870 and 1898, and that the effect of higher temperature and lower precipitation on desert plants remains a matter of conjecture. In the future it may be possible to develop a reliable climatological record before 1898 from tree ring and growth studies. In time we will certainly know more than we now do about the specific response of specific desert plants to specific increments of change in temperature and soil moisture. The real test of the hypothesis will be how well it accords with this new information.

Cattle and climate, climate and cattle. Of the four general explanations for the vegetative and hydrologic changes of recent years—cattle, rodents, fire suppression, and climate—two have been rejected as major causes, and two retained. Can the relative importance of cattle and climate be defined more closely?

The probability that large-scale grazing in Sonora between 1700 and 1880 had no marked effects seems to indicate that grazing, in itself, was not enough to initiate the changes. The evidence from sites like MacDougal Crater and the Melisas Islands indicates, furthermore, that some of the changes have occurred where cattle have never been, and are, therefore, probably the result of climatic variation alone. On the basis of this reasoning climate has to be accorded the more important role of the two. But is climatic variation then a sufficient explanation in itself for all of the changes? Probably not.

The more one views the events that have been dealt with in this study, the more persistently the problem obtrudes of their relation to earlier vegetative history. To what extent is this ebb and flow of plant life part of the normal rhythm of the desert region? To what extent is it a unique event?

The answers to these questions are in many respects at the core of the problem. Apart from practical implications—whether in time the trends will reverse themselves; whether in the interest of improving his habitat man can initiate or accelerate a reversal—is the pertinence of the answers to the problem of cause.

For if the changes stem from climatic variation alone, as climatic conditions revert to what they used to be—and they inevitably will—the mesquite will disappear, the grasses will return, and the landscape will again assume an uncluttered aspect. With a return to cooler, moister conditions the paloverdes and the oaks may migrate downslope into their old habitats; conditions for saguaro establishment may improve, and the future again see the development of stands like the one now dying in Saguaro National Monument; the river channels may heal and a cycle of filling commence. The last of these events, the geological record shows, has happened in the past; the possibility of many of the others can be inferred without difficulty.

But there still remains a residuum of change that

may be irreversible, and it is centered about events in the old grassland, where grazing has been heaviest. Can cooler, wetter conditions eradicate mesquite and *Acacia vernicosa?* In the light of what we know at present about the relation between these tenacious plants and their environment, and about competition between them and the grasses, it is not easy to imagine that a combination of rainfall and temperature alone will suffice.

Our present knowledge may, of course, be at fault, and as it improves, a mechanism for climatic removal may become clear. But alternatively the possibility exists that the woody invasions are not part of a recurrent cycle associated with climate; that a combination of overgrazing and climatic stress brought them. The suspicion remains that the desert grassland, by and large, is a thing of the past and that, short of spraying them with diesel oil or uprooting them with a chain and bulldozers, the shrubs are here to stay. There is no evidence that the elimination of grazing can bring about their disappearance, once they have become established. There is no evidence that recurrent burning can eradicate adult mesquites, or that in a stand of the adult trees, grass can grow densely enough to carry fire frequently enough to kill off all replacement seedlings during the century or so that it takes the seed-bearing adults to live out their normal span of years and die.

More and more the conviction grows that we are not merely at one stage of a cycle repeating itself, but that the past eighty years have witnessed the evolution of a significant new vegetation put together out of the floristic remains of the old. There is no sign that the evolution has seen its full development yet, or is slackening its pace.

About cause, then, the best answer seems to be that the new vegetation—if one may call it that—has not arisen from climatic variation alone, but in response to the unique combination of climatic and cultural stress imposed by the events of the past eighty years; that climate and cattle have united to produce it.

The warmer, drier climate has probably tended to favor the physiological responses of some species and to discriminate against those of others. It has probably reduced soil moisture values beyond a critical point at the lower edge of some ranges. It has possibly had the effect of increasing the intensity and duration of the hot, arid foresummer beyond the capacity of some plants to endure. By weakening the grass cover, domestic grazing animals have reinforced the general tendency toward aridity. They have contributed to an imbalance between infiltration and runoff in favor of the latter. This imbalance, in turn, may have been the event that triggered arroyo cutting. Because of the weakened grass cover the establishment of shrubby species has been facilitated. At the same time, livestock have aided in the dissemination of shrubby species, and the increased aridity has favored plants with their rooting habits. Where fire has been suppressed—if fire has been suppressed relative to its old frequency—and where rodents and jack rabbits have multiplied—if rodents and jack rabbits have multiplied beyond their old numbers—other factors may have made local contributions to the complex chain of events.

For now, this seems to be the best summary consistent with the evidence. As we find out more about the past climate and vegetation of the desert region, and more about the basic behavior of its plants, a better answer will inevitably emerge, based less on speculation and more on fact. For now the changing mile must remain a good subject for debate—but an even better subject for study.

NOTES

INTRODUCTION

[1] Mann (1963a) has given a penetrating analysis of Arizona's failure to adapt to hydrologic realities. In a second and more general work (1963b) he ably disputes Walter Prescott Webb's contention that political and social institutions do, in fact, tend to undergo modification in arid regions.

CHAPTER I

[1] A climatic classification of Arizona by William D. Sellers, using Trewartha's modification of Köppen appears in Green 1962: 2. Shreve's delineation of the Sonoran Desert may be found as Map 1 in Shreve 1964.

[2] McDonald (1962: 4) has estimated the daily influx of water vapor into Arizona from the south during July to be about 2,000,000 ac. ft.

[3] The best climatological information about the little-known area between Sonoyta and the Gulf of California is contained in Turnage and Mallery (1941). Basing their figures on records seven to ten years in length from storage gages, these authors give 5.6 in. per year for the "Pinacate Plateau" and 4.1 in. per year from Tule Tank, two locations that flank the part of the Pinacate region shown in the plates. They list 3.9 in. per year for Libertad, and 3.6 in. per year for Punto Cirio. Sykes (1931) has described the network of storage gages from which these data were gathered. The Secretaría of Recursos Hidráulicos since 1948 has operated a climatological station at Puerto Peñasco, and their data through 1962 give a mean annual value of about 3 in. for that location (Hastings 1964b). Other SRH stations have been recently installed at Costa Rica (1958–60); Tajitos (1960–); El Bámori (1962–); La Unión (1958–); and Puerto Libertad (1960–), and will in time reveal a great deal more about this section of the Gulf Coast.

[4] Defined as the standard deviation of the precipitation amounts divided by the mean amount. More properly, since precipitation distributions are positively skewed: the root-mean-square-deviation divided by the mean. Ignoring the fact of nonnormality, the coefficient may be roughly defined as the percentage of the mean precipitation by which the actual precipitation may be expected to depart during about one-third of the time periods. For probability data for nonnormal cases, see Hastings (1965a).

[5] Again ignoring nonnormality.

[6] Defined as $[(|r_1 - r_2| + |r_2 - r_3| + \ldots |r_{n-1} - r_n|) /(n - 1)]/\bar{r}$ where r_n is monthly rainfall during the nth year, and $\bar{r}$ the mean monthly rainfall for the period from the 1st through the nth years.

[7] Similar maps are presented by Dorroh (1946) and McDonald (1956).

[8] The initial stages of the experiment have been described by Hastings (1961b).

[9] The difference between a "secular trend" and an "unusual occurrence" is not obvious, and may be only a matter of how frequently an extreme recurs. In the case of large perennials, the microclimates in which the seedlings and the adults reside are substantially different. Close to the ground it is hotter by day and colder by night, and a mature plant which has survived these rigors as a seedling may find that even the intensity of once-in-one-hundred-years events in the more elevated, adult environment does not exceed what it has already experienced. The same event, more intense near the ground, may extinguish all seedlings, but since the seed source has survived, this is a matter of small consequence and can be remedied during the following year. Thus, only a continuous series of rare events can ultimately diminish the adult population by depriving it, season after season, of replacements.

[10] Although as Shreve (1964: 9) points out, the climate is nowhere very maritime.

[11] Estimated from the Sonoita record (Hastings 1964b: 118).

[12] (Shreve 1934b: 379–80, Turnage and Hinckley 1938: 547). Gentry, the observer at Cedros for Turnage and Hinckley, and their authority for the statement about the southern limit of the frost zone, states elsewhere that "in the Thorn Forest frost occasionally occurs, and during the excessive cold wave of January 1937 frost struck into the Short-tree Forest region" (Gentry 1942: 15).

[13] These, plus calm nights, are the classical requirements for the formation of an inversion. However, Dickson (1958: 39) points out that "inversions probably are more common in regions of diversified relief than has before been realized and . . . they frequently occur when a moderate cloud cover and light winds are present." Shreve (1912) provides some data dealing with the variation of inversions from season to season.

[14] Conduction and convection are the processes usually given (e.g. Sutton 1953: 132). However, observations that minimum air temperatures under certain conditions are found not at the soil surface, but from one inch to one foot or more higher, have led to hypotheses that direct radiation losses by the air may also be involved (Lake 1956).

[15] The phrase is borrowed from Geiger (1957: 195). Some useful temperature profiles through an inversion layer are presented by Young (1921).

[16] Bisbee, which is too warm for its elevation, can probably be explained by another topographic anomaly. The town occupies a narrow strip along the bottom of a steep canyon flanked on either side by mountains that rise sharply to a height of about five hundred feet. The canyon falls too rapidly to permit the accumulation of cold air; radiation and counterradiation from the slopes on either side tend at night to warm the town, or more accurately, to depress its cooling rate. The same factor may operate in the case of Fort Grant, nestled at the foot of the massive Pinaleño range.

[17] Although in cases this extreme, moisture considerations apart from temperature are probably involved too. The example is from Shreve (1922).

[18] At Death Valley, Went and Westergaard found that creosote bush, given adequate rainfall, germinated at temperatures intermediate between those optimum for, on the one hand, summer annuals and, on the other, winter annuals. They found abundant *Larrea* seedlings after an October rain when minimum temperatures were 15°–16° C., but not after rains toward the end of August (26° C.) or November (8°–10° C.). The distribution of *Larrea* on the lower, cooler part of a bajada may be influenced by these germination requirements (Went and Westergaard 1949).

[19] A case in point is the January 1937 cold wave already referred to. The duration of freezing at both the hill and garden stations was 19 hrs.; at Summerhaven, 130 hrs. (Turnage and Hinckley 1938: 538–42).

[20] The data in *Arizona Climate* show, with two exceptions, that no station lying in the desert has ever recorded a day with a maximum temperature of 32° or less. One of the exceptions is Bartlett Dam, which had supposedly experienced such a day in October. Inspection revealed that one digit had been dropped from a punch-card entry, and that a maximum temperature of 100° had been entered as 10°. The *Arizona Climate* entry is, therefore, in error, and has been corrected in Figure 5. The second exception is Silverbell; the present station lies in the desert; however, the maximum of less than 32° was recorded at an earlier location four miles away, at a cooler site in a valley on the northwest side of the Silverbell Mountains. The Silverbell temperature record, therefore, is distinctly nonhomogeneous. See U. S. Department of Commerce, Weather Bureau (1956: 81).

[21] For what it is worth in the largely philosophical debate among ecologists over the relative importance of means and extremes in controlling plant distributions: Shreve's Sonoran Desert boundary also coincides precisely with a mean monthly isotherm of 64.2° F. If extremes, in the form of record lows, are plotted, *no* isotherm will fit. That for a record low of 8.5° F. comes closest; however, it throws Bagdad, Miami, Globe, and Santa Rita Experimental Range headquarters into the Sonoran Desert, and excludes Tucson, Maricopa, and Casa Grande National Monument. The mean minimum temperature for January—more or less a hybrid between mean and extreme—gives somewhat better results; the isotherm for 47.55° F. excludes only Wickenburg, Reno Ranger Station, and Aguila. This is not to say, of course, that the physiological processes of a saguaro respond in any magic fashion to a mean temperature of 64.2° F. It merely states a truism that is frequently overlooked by "extremists," and that, therefore, needs to be emphasized: a mean is more apt than an extreme to correlate well with a large variety of other temperature measures, at least one of which—like duration of freezing here—may have real physiological significance. A mean is therefore more likely than an extreme to be useful in defining a plant's distribution, even though the actual limiting temperature factor may be unknown, and may be neither a mean nor an extreme.

CHAPTER II

[1] This persistent concept of civilized man as the irrational

element in an otherwise balanced and orderly world has been ably examined by Carl Lotus Becker (1932).

[2] Figure 7 has been redrawn from Sauer (1934). For a discussion of the Sumas and their relation to the Apaches who later occupied the same area see Forbes (1957), who disagrees with Sauer's interpretation.

[3] Spicer estimates that the ranchería groups numbered about 150,000. These include, however, 30,000 Mayos who fell outside the boundaries of the Sonoran Desert, and perhaps 15,000 of the 25,000 Opatas. To the remaining 105,000 must be added Seris, Yumans, and a few Apaches to arrive at the number who inhabited the area within.

[4] Two studies that pertain to other parts of New Spain, but nevertheless illuminate the decline of the aboriginal population under the impact of Spanish culture are Cook and Simpson (1948); Cook (1940).

[5] Nentuig (1951: 36–52). The work is usually attributed to an anonymous author, but Pradeau (1953) has demonstrated conclusively that Nentuig wrote it, and there seems to be no reason to continue the anonymous attribution.

[6] Baegert (1952: 67–68). Russell (1908: 71) notes the custom among the Pimas on the Gila Indian Reservation and suggests that "There would seem to be some special value ascribed to [such seeds, apart from their nutritional function]."

[7] McGregor, Alcorn, and Olin (1962: 266). These authors estimate that an average saguaro bears "about 4 flowers per day over a 30-day period each year," and that "half of these flowers normally set."

[8] A density of 116 plants per acre has been observed by the authors near Redington, Arizona. The stand shown in Plate 61 has about 33 per acre. A stand with 6 per acre, the figure used here, would be very open, but the density has been deliberately assumed low in order to arrive at a maximum impact for food gathering.

[9] With the saguaro, 12,000 ac. would suffice to feed 100,-000 people if the stand were as dense as that at Redington, Arizona.

[10] Computed on the basis of 2000 seeds per fruit (McGregor *et al.* 1962: 266).

[11] This discussion must be slightly qualified by the observation that other Indian groups may have made thriftier use of cactus fruits than did the Seris, on whose improvident habits these calculations are based. Pfefferkorn (1949: 76, 200) observes that his Pimas made cakes of the fruits, which "keep for two or three months without spoiling at all." Russell (1908: 72) describes both syrup and dried balls made from the fruit. "The supply [of fresh fruit] is a large one and only industry is required to make it available throughout the entire year." In any event, one can inject a factor of two or even four into the calculations without affecting the conclusion.

[12] Di Peso (1953), who conducted an archaeological investigation at the Quíburi site, traces the peregrinations of the ranchería in detail.

[13] The list of those who have expounded the fire hypothesis is a long and distinguished one and dates at least back to 1819. Humphrey (1958) has compiled a bibliography of works dealing with it, and has discussed its development.

[14] It is obviously impossible to treat here the involved questions of pre-Spanish ecology, and no attempt has been made even to describe those cultures next preceding the ones found by the conquistadors. An extensive literature exists for the Hohokam of southern Arizona; Sauer and Brand (1931) have surveyed what they consider to be the counterpart of the Hohokam in northern Sonora.

[15] (Bolton and Marshall 1920). The route of Fray Marcos and Coronado is in dispute. Undreiner (1947) and Schroeder (1956) disagree with Bolton and believe that Coronado went

all the way down the San Pedro, crossed the Gila, thence to the Salt, and from it to Zuñi. See also Sauer (1932). Translations of the sources may be found in Hammond and Ray (1940).

[16] The Jesuit advance has been well studied and much written about; a score of studies deal with special phases of it. Decorme (1941) presents a general survey.

[17] There is general agreement that the Apaches appeared on the northern frontier about 1680–90. Whether their migration was linked to the Pueblo Revolt, or to pressures from the Comanches is a matter of dispute (Sauer 1934: 59, 74, 81; Schroeder 1956: 28–29; Spicer 1962: 230–33; Forbes 1958; Worchester 1941).

[18] Tamarón's figures agree substantially with those given by Nicolás de Lafora (1939). They are considerably under Teodoro de Croix's figures for 1781 (Thomas 1941: 133).

[19] A thorough historical study needs to be made of the colonization of Sonora. A great deal of attention has been paid to missionary activity, but almost none to secular growth. Hastings (1961a) has surveyed the more obvious aspects.

[20] Di Peso (1953) places the location at the old village of Quíburi.

[21] Thomas (1941 and 1932). Chapman (1916: 386 ff.) disagrees with Thomas' estimates of de Croix's abilities.

[22] The process of hybridization can be traced by comparing the statistics in Tamarón's report with those in two reports by Bishop Reyes (Tamarón 1937, Reyes 1938, Reyes 1958). Cook (1942: 501) estimates that mestizos comprised 49 per cent of the population of New Spain by 1793 when Revilla-Gigedo's census was completed.

[23] Hardy (1829), Velasco (1850), and Zuñiga (1835) give good contemporary descriptions of Sonora during this period. Berber (1958) presents a general history of the state.

[24] Velasco (1850: 238) says the Indian attacks recommenced in 1832.

[25] The brief history of the Mexican grants presented here is taken from Ray H. Mattison (1946). Much of the source material may be found in the Journal of Private Land Grants, 5 vols., General Land Office, Phoenix, Arizona; and in the court records for the individual cases as they came before the Court of Private Land Claims, which existed from 1891 to 1904, and the United States Supreme Court. The appropriate references to each grant may be found in Mattison's work.

[26] No mention has been made of the grants whose validity was not upheld: El Sopori; Arivaca; Los Nogales de Elias; Tres Alamos; San Pedro; Agua Prieta; Naidenibachi and Sta. Bárbara; El Paso de los Algodones; or the notorious Peralta-Reavis claim. The disposition of all the cases may be found in *Report of the Attorney General: 1904*.

[27] As attractive as the grants were to speculators during and following the 1870's, it seems safe to assume that the slightest evidence of earlier occupation was made the basis for a claim.

[28] See Mattison (1946: 289, 315); *Tombstone Daily Epitaph*, Feb. 16, 1886; "Report of Surveyor-General Hise," *Arizona Daily Star*, July 18, 1886.

[29] For other descriptions of San Bernardino see Bartlett (1854: II, 255–56), Powell (1931: 127), Clarke (1852: 77).

[30] (Bartlett 1854: I, 386). Bartlett mistakenly calls Sonoita "Calabazas"; it is clear from his description, however, that he is just west of Patagonia; later, when he visits the true Calabasas, he corrects his error (*ibid.*, II, 307–8).

[31] (Bartlett 1854: I, 396–97). An excellent description of the old buildings may be found in Tevis (1954: 76–77).

[32] In 1848 Couts (1961: 58–59) found Guébabi, Tubac, and Tumacacori inhabited, and in between Santa Cruz and Guébabi he passed "several deserted as well as inhabited ranches." In 1852 Bartlett (1854: II, 304 ff.) found Tubac inhabited, but notes that the preceding year it was deserted. Tumacacori and the establishments along the river all the way to Santa Cruz were abandoned.

Two years later Bell found Tubac inhabited, and a thriving establishment owned by Manuel Gándara at Calabasas as well: "I learn that the owner—the governor of Sonora—intends abandoning it . . . it has only been a few weeks since, that the Indians killed fifty head of sheep, and are continually driving off the stock" (Haley 1932: 311). Another account of Gándara's operation at Calabasas may be found in Senate Executive Document 207 (U. S. Congress 1880), where the involved history of the Tumacacori-Calabasas grants is traced in some detail.

[33] Golder, Bailey, and Smith (1928: 193). See also Journal, Bliss (1931); Journal, Jones (1931); Extracts, Bigler (1932).

CHAPTER III

[1] Carson (1926); Pattie (1905). Secondary accounts of fur-trapping in the region may be found in Lavender (1954); Cleland (1950); Lockwood (1929).

[2] Etz (1939). For other references to beaver in the San Pedro River in the late nineteenth century see the following manuscripts in the Arizona Pioneers' Historical Society, Tucson: Pool (1935); Ohnesorgen (1929); Boedecker (1930).

[3] Bieber (1938: 42). References to fish in the Santa Cruz River, the San Pedro River, and Sonoita Creek are abundant. For some of the later ones see *Arizona Daily Star*, August 20, 1886; July 17, 1887. *Arizona Weekly Star*, July 26, 1877. Spring (1902). *The Tombstone*, August 27, 1885. *Weekly Arizonan*, May 12, 1859.

[4] Bartlett (1854: I, 379–81). For another account by Lt.-Col. Graham, see U. S. Congress (1852: 35–36).

[5] Compare Bryan (1925a); Cottam and Stewart (1940).

[6] Bartlett (1854: II, 324). Humphrey (1958: 211), misinterpreting Bartlett's route, states that "This open grassland gave way to mesquite as the party approached the San Pedro about seven miles south of the present International Boundary line." Mesquite was encountered two days journey east of the San Pedro, not in the marshy bottoms of that river. On the day he crossed the San Pedro Bartlett (1854: II, 325) stated: "It was with difficulty we could find scraps of wood enough to cook our dinner." The following day he commented that "scarcely a tree or bush had been seen since we left the vicinity of Santa Cruz. As the country continued bare today, the men picked up every fragment of wood or brush we passed." *Ibid.* Not until the day following this, two days after crossing the San Pedro, did he encounter the mesquite that Humphrey refers to. Humphrey is also in error about other parts of Bartlett's route. What he takes for "the San Rafael Valley in southeastern Arizona" is Cañada de la Zorilla in north central Sonora (Humphrey 1958: 211). Nor was the hacienda of San Bernardino located on "Black Draw in extreme southeastern Arizona" (*ibid.*, p. 204). It was located on San Bernardino Cienega near San Bernardino Creek, just south of the International Boundary in northeastern Sonora. San Bernardino Ranch presently occupies the location.

[7] A letter from James C. Malin, inquiring about an article in *Life*, August 18, 1952, dealing with the mesquite problem elicited the response that "a century and a half ago, there was hardly any [mesquite] in the U. S., but during the next fifty years it was brought into this country from Mexico by Spanish ponies and by wandering herds of wild buffalo . . . so that, in 1850 . . . scattered stands of mesquite were growing along the creeks and river beds of the southwest. This first generation mesquite, however, was exceedingly sparse" (Malin 1956: 449–50).

[8] Stewart (1951: 317–20); Branscomb (1956: 28–31). The question of fire on the midwestern prairies is an old and controversial one, not within the purview of this study. That the prairies represented a fire climax is disputed by many. See, for example, Weaver (1954: 271–73); Borchart (1950).

[9] Citations for the journals are as follows: Bieber (1938); Golder, Bailey, and Smith (1928); Journal, Bliss (1931); Extracts, Bigler (1932); Golder, Bailey, and Smith (1928); Couts (1961); Clarke (1852); Durivage (1937); Evans (1945); Cox (1925); Powell (1962); Aldrich (1950); Haley (1932); Emory (1848); Johnston (1848); Griffin (1943); Chamberlain (1945); Eccleston (1950); U. S. Congress (1859); Itinerary (1858); Bartlett (1854); U. S. Congress (1852).

[10] Branscomb presumably would disagree with this evaluation. In his study of shrub invasion on the Jornada of New Mexico he states that: "No reference to fires in the Jornada area has been found in the extensive review of historical literature connected with this particular study. This might lead to the conclusion that fire, or the lack of fire, has not been a factor in the encroachment of shrubby species upon grasslands in this area. On the other hand, it has been established that the periodic recurrence of wild fire was commonplace throughout the desert grasslands of the Southwest before white settlement. . . . Although no early travellers on the Jornada reported evidence of fire, the fact that fires were prevalent throughout the desert grassland and the area in question was being somehow maintained as desert grassland dominant, might well lead one to conclude that periodic fires, although not reported, did sweep the area and were a factor in restricting or preventing the spread of shrubs" (Branscomb 1956: 32–33). "Fire, which was started either by accident, by carelessness, or intentionally by Indians, acted as a prime factor in restricting shrub invasion on the desert grassland" (*ibid.*, p. 30). Where there is no smoke, there evidently must be fire.

[11] Among the best sources for recreating the Arizona scene in the 1870's are the lively, but frequently inaccurate contemporary handbooks: Hinton (1954); Hamilton (1881); *Hist. of A. T.* (1884).

[12] Several studies recount the history of ranching in the Gadsden Purchase: Haskett (1935); Morrisey (1950); Wagoner (1951); Wagoner (1952). The regional context is treated by Love (1916). The national picture is presented in such works as Osgood (1929); Dale (1930); Pelzer (1936).

[13] *Arizona Weekly Enterprise,* July 28, 1888. *Arizona Citizen,* September 9, 1889.

[14] *Arizona Daily Star,* August 5, 1890.

[15] *Ibid.,* August 6, 1890.

[16] *Ibid.,* August 7, 1890.

[17] *Ibid.,* August 8, 1890.

[18] *Ibid.,* August 9, 1890.

[19] *Ibid.,* August 13, 1890.

[20] Olmstead (1919: 79); Calvin (n.d.: 11, 39); Schwennesen (1917: 6); Swift (1926: 70–71).

[21] The Sonoyta River, which drains toward the Gulf of California, should not be confused with Sonoita Creek, a tributary of the Santa Cruz River.

[22] Rich (1911); Thornber (1910: 336–38); Cottam and Stewart (1940: 626); Winn (1926); Dittmer (1951).

[23] Leopold and Snyder (1951: 17). See also Leopold (1951b: 305).

[24] Elsewhere (Leopold *et al.* 1963) Leopold reiterates this explanation but suggests, like Martin (1963), that the arroyo cutting of the thirteenth century may not have been caused by aridity, but by a different balance of summer and winter precipitation.

[25] Reviews of the various positions may be found in Schumm and Hadley (1957: 161); Antevs (1952: 375–76).

CHAPTER IV

[1] Although there has been little quantitative ecology performed in the oak woodland, there exist a number of descriptive treatments. In addition to those already referred to are: Darrow (1944), Wallmo (1955).

[2] Martin (1963: 67) thinks that the isolation of Mexican montane elements like *Quercus emoryi* has taken place since "the mid-postglacial, 4,000 to 8,000 B.P., when the distinctive summer wet climate of the Mexican Plateau was especially well developed." In view of the fact that the lower border of the woodland has risen several hundred feet in recent years—a fact he is evidently unaware of—his statement (*ibid.,* 68), "I find no reliable pollen evidence that postglacial droughts, if they occurred, were sufficient to shift biotic zones above their present level," is probably more equivocal than he intended it to be. The present level of the lower edge of the woodland is considerably higher than it was 80 years ago, probably as the result of drought. A corresponding level at any time in the postglacial period would indicate an equally intense period of stress.

CHAPTER V

[1] The interested reader may refer to several general discussions: Clements (1920), Brand (1936), White (1948), Shantz (1924), Shreve (1917b, 1942b, 1942c), Humphrey (1958); or to some detailed studies of the grassland of the desert region: Wallmo (1955), Darrow (1944), Johnson (1961).

[2] Careful scrutiny of the vicinity by W. A. Sheffer has recently disclosed one living oak tree—an Emory oak—among many carcasses belonging to the same genus. The USDA Forest Products Laboratory at Madison, Wisconsin, has shown the wood from one of these carcasses to belong to a member of the red oak group, but identification to species is not possible (personal communication, February 12, 1963). Of the two red oaks in this part of Arizona (*Quercus hypoleucoides* being the other), only Emory oak can be expected at this elevation.

[3] Shreve (1942a: 235) recognized the changes from grassland to Chihuahuan Desert that have recently altered the appearance of the landscape in the San Pedro Valley. He assigned the area to grassland, an interpretation more recently followed by Martin (1963). Benson and Darrow (1954) described the same vegetation as Chihuahuan Desert without raising the question of recent change; this area is shown on their map, however, as Sonoran Desert. Nichol (1952) described the tarbush—creosote bush desert of this area but did not point out its resemblance to the Chihuahuan Desert.

CHAPTER VI

[1] Accounts of the 1907 expedition on which the Pinacate pictures were taken may be found in MacDougal (1908) and Hornaday (1908). Modern descriptions of the crater appear in Galbraith (1959) and Jahns (1959).

[2] For an example of fluctuation within a stand of jumping cholla see Tschirley and Wagle (1964). Thinning of the stand was attributed to a pathogen, possibly *Erwinia carnegieana.*

CHAPTER VII

[1] In the Tombstone vicinity (Plate 52), a group of investigators found that with increasing distance from the town, shrub cover (*Acacia vernicosa* and other woody species) diminished on soils of the Tortuga series and grass cover increased on soils of the Earp series. They suggested that the areas nearest town have received use and abuse longer than outlying districts, implying that cultural forces exercise a close control over present vegetation composition. Until more complete data are presented, this hypothesis cannot be ac-

cepted to the exclusion of other plausible explanations. For example, within the area near Tombstone occupied by the Earp series, elevation increases with distance from the town. It is likely that precipitation also increases along the slope, which rises almost 1000 ft. Thus either a rainfall gradient or man's activities could explain the observed differences in grass cover (Southwest Watershed 1958).

CHAPTER VIII

[1] Some recent discussions of desertification appear in Whyte (1963), Huxley (1961), Huzayyin (1956), Man's Tenure (1956), Talbot (1961), Stamp (1961), Forbes (1958). Sears (1947) has written an eloquent, but not notably objective tract on the subject.

[2] For an amusing example of smoothing-induced cycles see Cole (1957).

[3] Partial listings of the extensive bibliography that has grown up in connection with the fluctuation may be found in Meteorological Abstracts (1950), Shapley (1953), and Mitchell (1961). A comprehensive discussion by Veryard (1963) of the literature appears in the Proceedings of the Rome Symposium on Climatic Change. Many of the papers presented at the Conference on Solar Variations, Climatic Change and Related Geophysical Problems (New York, 1961) bear on the topic, and may be found in Furness (1961).

[4] For a comprehensive discussion of urbanization and its relation to temperature increase see Mitchell (1953).

[5] The records used are those of Ajo, Ashfork, Bartlett Dam, Douglas, Dudleyville, Flagstaff, Florence, Grand Canyon, Natural Bridge, Pinedale, Roosevelt, Sacaton, Salome, Sierra Ancha, Tombstone, Window Rock, Yuma, and Yuma Citrus. No known station moves were encompassed by the analysis. The records are of varying lengths; the difficulty of averaging them was surmounted by converting the data into departures from the mean temperature for the station for the period 1936–45.

[6] Dr. Sellers has very kindly made available his calculations. In his original article, of course (Sellers 1960), the data were not smoothed by ten-year running means.

[7] Martin (1963: 4) presents a general climatological model that, if applied to the recent past, is incompatible with this hypothesis. Basing his view on work by Wallén (1955), McDonald (1956), Sellers (1960), and personal correspondence with McDonald, he maintains that high summer temperatures in the American Midwest are associated with increased summer rainfall in the Southwest; that conversely cool summers in the Midwest are associated with drought in Mexico and the Southwest. Since he asserts elsewhere (*ibid.,*

8, 67) that with increased summer rainfall the oak woodland migrates to lower elevations, it is clear that he associates three events: (1) increased temperatures in the Midwest; (2) increased summer rains in the Southwest; (3) the downward migration of the oak woodland if the first two phenomena are marked and persist.

Our preceding chapters have shown that in the specific situation since 1898, temperatures in the Midwest (and Southwest) have been going up, that summer rains in the Southwest have remained about the same or have gone down slightly, and that the oak woodland has migrated upward.

So far as we can see, the evidence for the upward movement of the woodland in recent years is beyond question; so far as we can see, there is no similarity at all between Seller's curve for summer rainfall in Arizona and New Mexico (Figure 14) and Kincer's curves for Midwestern temperature (Figure 13). In the period since 1898 Midwestern temperatures have risen; Southwestern summer rainfall has tended to remain about the same or to decrease slightly. We are frankly puzzled at what in the work of McDonald (1956) or Wallén (1955) (see Martin 1963: 67) would lead Martin to conclude that summer rainfall has gone up.

The larger question involved, whether Martin's climatological model is incorrect, or alternatively, whether it is correct for Altithermal time where he employs it, but not for recent times, will not be considered. The model does not, at any rate, accord with the facts of climate since 1898.

CHAPTER IX

[1] Humphrey (1962: 165–67) gives a review of the subject.

[2] Phillips (1912), Humphrey (1962: 167), Humphrey (1960: 19). In the case of *Quercus turbinella, Q. dumosa* and other scrub species of the genus, fire may result in multiple sprouting from the root, with the eventual emergence, after repeated burning, of a distinctive chaparral vegetation. Phillips (1912: 10) notes such a tendency in Emory oak also. Although the discussion here is concerned with the oak savannas, shown in the old plates, not the oak chaparral, the two types cannot always be divorced easily. Benson (1962: 70–71) gives an account of the effect of fire in California on a border where oak woodland (*Q. douglasii*) and oak chaparral (*Q. dumosa*) meet.

[3] Parker and Martin (1952: 15), Humphrey (1949: 175–82), Reynolds and Bohning (1956). The three studies more or less span opinion on the effectiveness of burning in controlling mesquite. To the writers' knowledge no studies describe the effect of burning on a complex where oaks and mesquite grow side by side. This comparative information would be most valuable.

APPENDIX A

Scientific Equivalents for Common Names
Used in the Text and Captions

agave *Agave* spp.
alligator juniper *Juniperus deppeana* Steud.
all thorn *Koeberlinia spinosa* Zucc. var. *spinosa.*
amole *Agave schottii* Engelm.
Arizona madrone *Arbutus arizonica* (Gray) Sarg.
Arizona rosewood *Vauquelinia californica* (Torr.) Sarg.
Arizona sycamore *Platanus wrightii* Wats.
Arizona white oak *Quercus arizonica* Sarg.
barrel cactus *Ferocactus* spp.
beardgrass *Andropogon glomeratus* (Walt.) B.S.P.
beargrass *Nolina microcarpa* Wats.
Bermuda grass *Cynodon dactylon* (L.) Pers.
big galleta *Hilaria rigida* (Thurb.) Benth.
bisnaga *Ferocactus wislizeni* (Engelm.) Britt. & Rose.
black grama *Bouteloua eriopoda* Torr.
black mangrove *Avicennia nitida* Jacq.
blue grama *Bouteloua gracilis* (H.B.K.) Lag.
blue paloverde *Cercidium floridum* Benth.
blue stem *Andropogon* spp.
blue yucca *Yucca baccata* Torr. var. *brevifolia* (Schott) Benson & Darrow.
boojum tree *Idria columnaris* Kellogg.
bristlegrass *Setaria macrostachya* H.B.K.
brittle bush *Encelia farinosa* Gray.
bullnettle *Solanum elaeagnifolium* Cav.
bursage *Franseria deltoidea* Torr.
burrobrush *Hymenoclea monogyra* Torr. & Gray.
burroweed *Haplopappus tenuisectus* (Greene) Blake.
bush muhly *Muhlenbergia porteri* Scribn.
cane cholla *Opuntia spinosior* (Engelm.) Toumey.
canyon grape *Vitis arizonica* Engelm.
canyon ragweed *Franseria ambrosioides* Cav.
cardón *Pachycereus pringlei* (Wats.) Britt. & Rose.
catclaw *Acacia greggii* Gray.
century plant *Agave* spp.
chamiso *Atriplex canescens* (Pursh) Nutt.
cholla *Opuntia* spp. (*Cylindropuntias*).
chuparosa *Beloperone californica* Benth.

cliff rose *Cowania mexicana* D. Don. var. *stansburiana* (Torr.) Jepson.
cockroach plant *Haplophyton crooksii* L. Benson.
coffeeberry *Rhamnus californica* Esch.
cottonwood *Populus fremontii* Wats.
creosote bush *Larrea tridentata* (DC.) Coville.
crownbeard *Verbesina encelioides* (Cav.) Benth. & Hook.
curly mesquite *Hilaria belangeri* (Steud.) Nash.
desert broom *Baccharis sarothroides* Gray.
desert cotton *Gossypium thurberi* Todaro.
desert hackberry *Celtis pallida* Torr.
desert holly *Perezia nana* Gray.
desert honeysuckle *Anisacanthus thurberi* (Torr.) Gray.
desert lavender *Hyptis emoryi* Torr.
desert mallow *Sphaeralcea* (*ambigua* Gray)?
desert saltbush *Atriplex polycarpa* (Torr.) Wats.
desert willow *Chilopsis linearis* (Cav.) Sweet.
desert zinnia *Zinnia pumila* Gray.
Douglas fir *Pseudotsuga menziesii* (Mirbel) Franco.
doveweed *Croton texensis* (Klotzsch) Muell. Arg.
elephant tree *Bursera microphylla* Gray.
Emory oak *Quercus emoryi* Torr.
Engelmann prickly pear *Opuntia engelmannii* Salm.-Dyck.
fairyduster *Calliandra eriophylla* Benth.
fluffgrass *Tridens pulchellus* (H.B.K.) Hitchc.
foothill paloverde *Cercidium microphyllum* (Torr.) Rose & Johnston.
giant cactus (saguaro) *Carnegiea gigantea* (Engelm.) Britt. & Rose.
Goodding willow *Salix gooddingii* Ball.
gray thorn *Condalia lycioides* (Gray) Weberb.
grama grass *Bouteloua* spp.
hairy grama *Bouteloua hirsuta* Lag.
hopbush *Dodonaea viscosa* var. *angustifolia* (L.F.) Benth.
ironwood *Olneya tesota* Gray.

jojoba   *Simmondsia chinensis* (Link) Schneid.
Joshua tree   *Yucca brevifolia* Engelm.
jumping cholla   *Opuntia fulgida* Engelm.
juniper   *Juniperus* spp.
kidneywood   *Eysenhardtia polystachya* (Ortega) Sarg.
limber bush   *Jatropha cardiophylla* (Torr.) Muell. Arg.
loosestrife   *Lythrum californicum* Torr. & Gray.
mala mujer   *Cnidoscolus angustidens* Torr.
mangrove   *Rhizophora mangle* L., *Avicennia nitida* Jacq.
mariola   *Parthenium incanum* H.B.K.
mescal   *Agave palmeri* Engelm.
mesquite   *Prosopis juliflora* (Swartz) DC. var. *velutina* (Woot.) Sarg.
Mexican blue oak   *Quercus oblongifolia* Torr.
Mexican crucillo   *Condalia spathulata* (Gray) Weberb.
Mexican devilweed   *Aster spinosus* Benth.
Mexican pinyon pine   *Pinus cembroides* Zucc.
Mexican tea   *Ephedra trifurca* Torr.
mimosa   *Mimosa* spp.
mortonia   *Mortonia scabrella* Gray.
mountain mahogany   *Cercocarpus breviflorus* Gray.
mountain yucca   *Yucca schottii* Engelm.
netleaf hackberry   *Celtis reticulata* Torr.
netleaf oak   *Quercus rugosa* Née.
oak   *Quercus* spp.
ocotillo   *Fouquieria splendens* Engelm.
one-seed juniper   *Juniperus monosperma* (Engelm.) Sarg.
organpipe cactus   *Lemaireocereus thurberi* (Engelm.) Britt. & Rose.
palmilla   *Yucca elata* Engelm.
paloverde   *Cercidium* spp.
paperdaisy   *Psilostrophe cooperi* (Gray) Greene.
pencil cholla   *Opuntia arbuscula* Engelm.
pigmy cedar   *Peucephyllum schottii* Gray.
pincushion cactus   *Mammillaria microcarpa* Engelm.
pine   *Pinus* spp.
pointleaf manzanita   *Arctostaphylos pungens* H.B.K.
poison ivy   *Rhus radicans* L.
ponderosa pine   *Pinus ponderosa* Laws.
prickly pear   *Opuntia* spp. (*Platyopuntias*).
rabbitbrush   *Chrysothamnus nauseosus* (Pall.) Britton var. *latisquameus* (Gray) H. M. Hall.
range ratany   *Krameria parvifolia* Benth.
red mangrove   *Rhizophora mangle* L.
red willow   *Salix laevigata* Bebb.
Rothrock grama   *Bouteloua rothrockii* Vasey.
Russian thistle   *Salsola kali* L. var. *tenuifolia* Tausch.
sacaton   *Sporobolus airoides* Torr.
saguaro (giant cactus)   *Carnegiea gigantea* (Engelm.) Britt. & Rose.

saltbush   *Atriplex* spp.
Santa-Rita cactus   *Opuntia santa-rita* (Griffiths & Hare) Rose.
seep willow   *Baccharis glutinosa* Pers.
senna   *Cassia covesii* Gray.
shrubby senna   *Cassia wislizeni* Gray.
sideoats grama   *Bouteloua curtipendula* (Michx.) Torr.
silverleaf oak   *Quercus hypoleucoides* Camus.
sinita   *Lophocereus schottii* (Engelm.) Britt. & Rose.
skunkbush   *Rhus trilobata* Nutt.
slender grama   *Bouteloua filiformis* (Fourn.) Griffiths.
smoketree   *Dalea spinosa* Gray.
snakeweed   *Gutierrezia lucida* Greene.
soapberry   *Sapindus saponaria* L. var. *drummondii* (Hook. & Arn.) L. Benson.
sotol   *Dasylirion wheeleri* Wats.
spruce   *Picea* spp.
sprucetop grama   *Bouteloua chondrosioides* (H.B.K.) Benth.
staghorn cholla   *Opuntia versicolor* Engelm.
stickleaf   *Mentzelia* spp.
stickweed   *Stephanomeria pauciflora* (Torr.) A. Nels.
tanglehead   *Heteropogon contortus* (L.) Beauv.
tarbush   *Flourensia cernua* DC.
teddybear cholla   *Opuntia bigelovii* Engelm.
Texas mulberry   *Morus microphylla* Buckl.
threadleaf groundsel   *Senecio longilobus* Benth.
three-awn   *Aristida* spp.
tobosa   *Hilaria mutica* (Buckl.) Benth.
Toumey oak   *Quercus toumeyi* Sarg.
turpentine bush   *Haplopappus laricifolius* Gray.
velvet ash   *Fraxinus pennsylvanica* Marshall subsp. *velutina* (Torrey) G. N. Miller.
vine mesquite   *Panicum obtusum* H.B.K.
wait-a-minute   *Mimosa biuncifera* Benth.
walnut   *Juglans major* (Torr.) Heller.
watercress   *Rorippa nasturtium-aquaticum* (L.) Schinz & Thell.
western coral bean   *Erythrina flabelliformis* Kearney.
white bursage   *Franseria dumosa* Gray.
white fir   *Abies concolor* (Gordon & Glendinning) Hoopes.
white ratany   *Krameria grayi* Rose & Palmer.
white thorn   *Acacia constricta* Benth.
wild buckwheat   *Eriogonum abertianum* Torr.
Wright lippia   *Aloysia wrightii* (Gray) Heller.
Wright's silktassel   *Garrya wrightii* Torr.
yerba de pasmo   *Baccharis pteronioides* DC.
yewleaf willow   *Salix taxifolia* H.B.K.
yucca   *Yucca* spp.

APPENDIX B

Common Equivalents Used in the Text and Captions
for Scientific Names

Abies concolor (Gordon & Glendinning) Hoopes. white fir

Acacia constricta Benth.   white thorn

Acacia greggii Gray.   catclaw

Acacia vernicosa Standl.   none

Agave palmeri Engelm.   mescal

Agave schottii Engelm.   amole

Agave spp.   century plant

Aloysia wrightii (Gray) Heller.   Wright lippia

Amoreuxia palmatifida Moç. & Sessé.   none

Andropogon glomeratus (Walt.) B.S.P.   beardgrass

Andropogon spp.   blue stem

Anisacanthus thurberi (Torr.) Gray.   desert honey-suckle

Arbutus arizonica (Gray) Sarg.   Arizona madrone

Arctostaphylos pungens H.B.K.   pointleaf manzanita

Aristida glauca (Nees) Walp.   none

Aristida spp.   three-awn

Aster spinosus Benth.   Mexican devilweed

Atamisquea emarginata Miers.   none

Atriplex barclayana (Benth.) Dietr.   none

Atriplex canescens (Pursh) Nutt.   chamiso

Atriplex polycarpa (Torr.) Wats.   desert saltbush

Atriplex spp.   saltbush

Avicennia nitida Jacq.   black mangrove

Ayenia pusilla L.   none

Baccharis glutinosa Pers.   seep willow

Baccharis neglecta Britton.   none

Baccharis pteronioides DC.   *yerba de pasmo*

Baccharis sarothroides Gray.   desert broom

Bahia absinthifolia Benth.   none

Beloperone californica Benth.   *chuparosa*

Bouteloua chondrosioides (H.B.K.) Benth.   sprucetop grama

Bouteloua curtipendula (Michx.) Torr.   sideoats grama

Bouteloua eriopoda Torr.   black grama

Bouteloua filiformis (Fourn.) Griffiths.   slender grama

Bouteloua gracilis (H.B.K.) Lag.   blue grama

Bouteloua hirsuta Lag.   hairy grama

Bouteloua rothrockii Vasey.   Rothrock grama

Bouteloua spp.   grama grass

Brickellia spp.   none

Bursera hindsiana (Benth.) Engler.   none

Bursera microphylla Gray.   elephant tree

Calliandra eriophylla Benth.   fairyduster

Carlowrightia [*linearifolia* (Torr.) Gray]?   none

Carnegiea gigantea (Engelm.) Britt. & Rose.   saguaro (giant cactus)

Cassia covesii Gray.   senna

Cassia wislizeni Gray.   shrubby senna

Ceanothus greggii Gray.   none

Celtis pallida Torr.   desert hackberry

Celtis reticulata Torr.   netleaf hackberry

Cercidium floridum Benth.   blue paloverde

Cercidium microphyllum (Torr.) Rose & Johnston. foothill paloverde

Cercidium spp.   paloverde

Cercocarpus breviflorus Gray.   mountain mahogany

Chilopsis linearis (Cav.) Sweet.   desert willow

Chrysothamnus nauseosus (Pall.) Britton var. *latisquameus* (Gray) H. M. Hall.   rabbitbrush

Cnidoscolus angustidens Torr.   *mala mujer*

Coldenia canescens DC.   none

Condalia lycioides (Gray) Weberb.   gray thorn

Condalia spathulata (Gray) Weberb.   Mexican crucillo

Coursetia microphylla Gray.   none

Cowania mexicana D. Don. var. *stansburiana* (Torr.) Jepson.   cliff rose

Croton texensis (Klotzsch) Muell. Arg.   doveweed

Cynodon dactylon (L.) Pers.   Bermuda grass

Dalea spinosa Gray.   smoketree

Dasylirion wheeleri Wats.   sotol

Dodonaea viscosa var. *angustifolia* (L.F.) Benth.   hopbush

Dyssodia porophylloides Gray.   none

Echeveria pulverulenta Nutt.   none

Encelia farinosa Gray.   brittle bush

Ephedra trifurca Torr.   Mexican tea

Eriogonum abertianum Torr.   wild buckwheat
Erythrina flabelliformis Kearney.   western coral bean
Euphorbia misera Benth.   none
Euphorbia tomentulosa Wats.   none
Eysenhardtia polystachya (Ortega) Sarg.   kidneywood
Fagonia californica Benth.   none
Ferocactus spp.   barrel cactus
Ferocactus wislizeni (Engelm.) Britt. & Rose.   bisnaga
Flourensia cernua DC.   tarbush
Fouauieria splendens Engelm.   ocotillo
Frankenia palmeri Wats.   none
Franseria ambrosioides Cav.   canyon ragweed
Franseria confertiflora (DC.) Rydb.   none
Franseria deltoidea Torr.   bursage
Franseria dumosa Gray.   white bursage
Fraxinus pennsylvanica Marshall subsp. *velutina* (Torr.)
   G. N. Miller.   velvet ash
Garrya wrightii Torr.   Wright's silktassel
Gossypium thurberi Todaro.   desert cotton
Gutierrezia lucida Greene.   snakeweed
Haplopappus laricifolius Gray.   turpentine bush
Haplopappus tenuisectus (Greene) Blake.   burroweed
Haplophyton crooksii L. Benson.   cockroach plant
Heteropogon contortus (L.) Beauv.   tanglehead
Hibiscus coulteri Harv.   none
Hibiscus denudatus Benth.   none
Hilaria belangeri (Steud.) Nash.   curly mesquite
Hilaria mutica (Buckl.) Benth.   tobosa
Hilaria rigida (Thurb.) Benth.   big galleta
Hymenoclea monogyra Torr. & Gray.   burrobrush
Hyptis emoryi Torr.   desert lavender
Idria columnaris Kellogg.   boojum tree
Janusia gracilis Gray.   none
Jatropha cardiophylla (Torr.) Muell. Arg.   limber bush
Jatropha cinerea (Ortega) Muell. Arg.   none
Jatropha cuneata Wiggins & Rollins.   none
Juglans major (Torr.) Heller.   walnut
Juniperus deppeana Steud.   alligator juniper
Juniperus monosperma (Engelm.) Sarg.   one-seed juni-
   per
Juniperus spp.   juniper
Krameria grayi Rose & Palmer.   white ratany
Krameria parvifolia Benth.   range ratany
Koeberlinia spinosa Zucc. var. *spinosa*.   all thorn
Larrea tridentata (DC.) Coville.   creosote bush
Lemaireocereus thurberi (Engelm.) Britt. & Rose.   or-
   ganpipe cactus
Lophocereus schottii (Engelm.) Britt. & Rose.   *sinita*
Lycium berlandieri Dunal.   none
Lycium macrodon Gray.   none
Lycium pallidum Miers.   none
Lythrum californicum Torr. & Gray.   loosestrife
Mammillaria microcarpa Engelm.   pincushion cactus
Mentzelia spp.   stickleaf
Mimosa biuncifera Benth.   wait-a-minute
Mimosa dysocarpa Benth.   none
Mimosa grahami Gray.   none
Mimosa spp.   mimosa
Mortonia scabrella Gray.   mortonia
Morus microphylla Buckl.   Texas mulberry
Muhlenbergia porteri Scribn.   bush muhly
Nolina microcarpa Wats.   beargrass
Olneya tesota Gray.   ironwood

Opuntia arbuscula Engelm.   pencil cholla
Opuntia bigelovii Engelm.   teddybear cholla
Opuntia chlorotica Engelm. & Bigel.   none
Opuntia engelmannii Salm.-Dyck.   Engelmann prickly
   pear
Opuntia fulgida Engelm.   jumping cholla
Opuntia phaeacantha Engelm.   none
Opuntia santa-rita (Griffiths & Hare) Rose.   Santa-Rita
   cactus
Opuntia spinosior (Engelm.) Toumey.   cane cholla
Opuntia spp. (*Cylindropuntias*).   cholla
Opuntia spp. (*Platyopuntias*).   prickly pear
Opuntia stanlyi Engelm.   none
Opuntia versicolor Engelm.   staghorn cholla
Pachycereus pringlei (Wats.) Britt. & Rose.   *cardón*
Panicum obtusum H.B.K.   vine mesquite
Parthenium incanum H.B.K.   mariola
Perezia nana Gray.   desert holly
Peucephyllum schottii Gray.   pigmy cedar
Picea spp.   spruce
Pinus cembroides Zucc.   Mexican pinyon pine
Pinus ponderosa Laws.   ponderosa pine
Pinus spp.   pine
Platanus wrightii Wats.   Arizona sycamore
Populus fremontii Wats.   cottonwood
Prosopis juliflora (Swartz) DC. var. *velutina* (Woot.)
   Sarg.   mesquite
Pseudotsuga menziesii (Mirbel) Franco.   Douglas fir
Psilostrophe cooperi (Gray) Greene.   paperdaisy
Quercus arizonica Sarg.   Arizona white oak
Quercus emoryi Torr.   Emory oak
Quercus hypoleucoides Camus.   silverleaf oak
Quercus oblongifolia Torr.   Mexican blue oak
Quercus rugosa Née.   netleaf oak
Quercus spp.   oak
Quercus toumeyi Sarg.   Toumey oak
Rhamnus californica Esch.   coffeeberry
Rhizophora mangle L.   red mangrove
Rhus choriophylla Woot. & Standl.   none
Rhus microphylla Engelm.   none
Rhus radicans L.   poison ivy
Rhus trilobata Nutt.   skunkbush
Rorippa nasturtium-aquaticum (L.) Schinz. & Thell.
   watercress
Sageretia wrightii Wats.   none
Salix gooddingii Ball.   Goodding willow
Salix laevigata Bebb.   red willow
Salix taxifolia H.B.K.   yewleaf willow
Salsola kali L. var. *tenuifolia* Tausch.   Russian thistle
Sapindus saponaria L. var. *drummondii* (Hook. & Arn.)
   L. Benson.   soapberry
Senecio longilobus Benth.   threadleaf groundsel
Setaria macrostachya H.B.K.   bristlegrass
Simmondsia chinensis (Link) Schneid.   jojoba
Solanum elaeagnifolium Cav.   bullnettle
Solanum hindsianum Benth.   none
Sphaeralcea (*ambigua* Gray)?   desert mallow
Sporobolus airoides Torr.   sacaton
Stephanomeria pauciflora (Torr.) A. Nels.   stickweed
Tridens pulchellus (H.B.K.) Hitchc.   fluffgrass
Vauquelinia californica (Torr.) Sarg.   Arizona rose-
   wood

Verbesina encelioides (Cav.) Benth. & Hook.   Crown-beard

Vitis arizonica Engelm.   canyon grape

Yucca arizonica McKelvey.   none

Yucca baccata Torr. var. *brevifolia* (Schott) Benson & Darrow.   blue yucca

Yucca brevifolia Engelm.   Joshua tree

Yucca elata Engelm.   palmilla

Yucca schottii Engelm.   mountain yucca

Yucca spp.   yucca

Zinnia pumila Gray.   desert zinnia

Zizyphus sonorensis Wats.   none

APPENDIX C

The Location and Elevation of Photographic Stations; Photo Credits

PLATE	VICINITY	S.	T.	R.	ELEV.	CREDIT	PLATE	VICINITY	S.	T.	R.	ELEV.	CREDIT
		LOCATION							LOCATION				
	OAK WOODLAND						31	Helvetia	23	18	15	4300	A
1	El Plomo Mine	13	21	14	5500	A	32	Helvetia	23	18	15	4300	A
2	El Plomo Mine	13	21	14	5500	A	33	Hacienda Santa					
3	El Plomo Mine	13	21	14	5500	A		Rita	22	21	14	4150	A
4	Barrel Canyon	33	18	16	5100	B	34	Proto Ridge	2	24	14	4100	A
5	Guajalote Peak	[23]	23	15	4750	A	35	Davidson Canyon	1	18	16	4100	M
6	Sonoita Creek	3	21	16	4750	C	36	Davidson Canyon	1	18	16	4050	M
7	Sonoita Creek	33	20	16	4600	C	37	Davidson Canyon	1	18	16	4050	M
8	Fort Crittenden	27	20	16	4700	C	38	Elephant Head	13	19	14	3900	M
9	Monkey Spring	3	21	16	4750	C	39	Cerro Colorado					
10	Monkey Spring	3	21	16	4600	A		Mine	25	20	10	3650	A
11	Monkey Spring	3	21	16	4600	A	40	Tumacacori					
12	Monkey Spring	3	21	16	4550	A		Mission	30	21	13	3350	A
13	Alto Gulch	14	21	14	4650	A	41	Carmen	19	21	13	3250	A
14	Salero Mine	25	21	15	4500	A							
15	South of Nogales	—	—	—	4300	S		**Grassland of the San Pedro**					
16	San Cayetano Hill	32	22	15	4200	A	42	The Mexican Boundary	24	24	21	4350	A
17	San Cayetano Hill	32	22	15	4200	A	43	The Mexican Boundary	—	—	—	4350	A
18	Florida Canyon	19	19	15	4100	M	44	Greene Ranch	4	24	22	4250	A
19	Faber Canyon	13	19	14	3950	M	45	Spring Creek	24	23	22	4400	A
20	Proto Ridge	2	24	14	4000	A	46	Above Lewis Springs	[32]	21	22	4050	A
21	Proto Ridge	2	24	14	4000	A	47	Above Lewis Springs	[32]	21	22	4050	A
22	Buena Vista Grant	[6]	24	15	3950	A	48	Escapule Ranch	19	21	22	4050	A
23	Buena Vista Grant	[6]	24	15	3900	A	49	Gird Dam	12	21	21	4000	A
24	Buena Vista Grant	[6]	24	15	3800	A	50	Charleston	[1]	20	21	4050	A
25	Near Buena Vista	31	23	15	3800	A	51	Charleston	[1]	20	21	4000	A
26	Querobabi, Sonora	—	—	—	3650	S	52	Tombstone	14	20	22	4850	A
							53	Walnut Gulch	1	20	21	4000	A
	DESERT GRASSLAND						54	Walnut Gulch	1	20	21	4000	A
							55	Walnut Gulch	1	20	21	4000	A
	Grassland of the Santa Ritas						56	Walnut Gulch	2	20	21	3950	A
27	Red Rock Canyon	4	22	16	4450	C	57	Babocomari Creek	[4]	20	21	3800	A
28	Red Rock Canyon	4	22	16	4450	C	58	Above St. David	18	18	21	3700	A
29	Helvetia	23	18	15	4400	A	59	Above St. David	18	18	21	3700	A
30	Helvetia	23	18	15	4300	A	60	Above St. David	19	18	21	3700	A

		LOCATION							LOCATION				
PLATE	VICINITY	S.	T.	R.	ELEV.	CREDIT	PLATE	VICINITY	S.	T.	R.	ELEV.	CREDIT

DESERT

Arizona Uplands

61	Saguaro Nat'l Mon't	20	14	16	3050	N
62	Saguaro Nat'l Mon't	20	14	16	2950	N
63	Saguaro Nat'l Mon't	20	14	16	2950	S
64	Saguaro Nat'l Mon't	20	14	16	2950	S
65	Saguaro Nat'l Mon't	20	14	16	2900	S
66	Near Soldier Canyon	18	13	16	2950	M
67	Near Soldier Canyon	18	13	16	2900	M
68	Near Soldier Canyon	18	13	16	2900	M
69	Near Soldier Canyon	18	13	16	2850	M
70	Sabino Canyon	9	13	15	2850	A
71	Sabino Canyon	9	13	15	2850	A
72	Sabino Canyon	9	13	15	2850	A
73	Cañada del Oro	5	11	14	2650	M
74	Cañada del Oro	5	11	14	2650	M
75	Robles Pass	31	14	13	2650	M
76	Cat Mountain	25	14	12	2750	M
77	Brown Mountain	21	14	12	2650	M
78	Saguaro N.M. West	17	13	12	2900	M
79	Saguaro N.M. West	17	13	12	2900	M
80	Picture Rocks	9	13	12	2400	M
81	Picture Rocks	9	13	12	2400	M

Lower Colorado Valley

82	MacDougal Pass	—	—	—	900	M
83	MacDougal Pass	—	—	—	900	M
84	Macias Pass	—	—	—	900	M
85	MacDougal Crater	—	—	—	900	M
86	MacDougal Crater	—	—	—	900	M
87	MacDougal Crater	—	—	—	400	—
88	MacDougal Crater	—	—	—	900	M

Central Gulf Coast

89	Near Libertad	—	—	—	450	S
90	Near Libertad	—	—	—	450	S
91	Near Libertad	—	—	—	200	S
92	Punto Cirio	—	—	—	150	S
93	Near Punto Cirio	—	—	—	100	S
94	Punto Cirio	—	—	—	50	S
95	Libertad	—	—	—	50	S
96	Islas Melisas	—	—	—	5	M
97	Islas Melisas	—	—	—	5	M

The location is given in terms of section (S.), township south (T.), and range east (R.), on the standard grid for Arizona. No equivalents exist for stations in Mexico. Brackets denote a section number that does not officially exist, but which has been extrapolated from the official system to facilitate designating the locations.

Photo credits are for the old half of each pair, and are designated as follows: A, Arizona Pioneers' Historical Society Collection; B, personal collection of Mr. Albert C. Stewart; C, personal collection of Miss Doris Seibold; M, D. T. MacDougal collection, presently in the custody of Dr. Charles H. Lowe, Jr.; N, National Park Service; S, H. L. Shantz collection, The University of Arizona Herbarium, courtesy of Dr. Walter S. Phillips.

LITERATURE CITED

Ahlmann, H. W. 1949. "The Present Climatic Fluctuation," The Geographical Journal, CXII, 165–95.

Alcorn, Stanley M. 1961. "Natural History of the Saguaro," Arid Lands Colloquia, 1959–60, 1960–61. The University of Arizona, pp. 23–29.

———, and May, Curtis. 1962. "Attrition of a Saguaro Forest," Plant Disease Reporter, XLVI, 156–58.

Aldous, A. E., and Shantz, H. L. 1924. "Types of Vegetation in the Semiarid Portion of the United States and Their Economic Significance," Journal of Agricultural Research, XXVIII, 99–127.

Aldrich, Lorenzo D. 1950. A Journal of the Overland Route to California and the Gold Mines. Los Angeles: Dawson's Book Shop.

Alegre, Francisco Javier. 1956–60. Historia de la Provincia de la Compañía de Jesús de Nueva España. 4 vols. Edited by Ernest J. Burrus and Felix Zubillaga. Rome: Institutum Historicum, S. J.

Allison, Warren. n.d. "Arizona the Last Frontier." MS in Arizona Pioneers' Historical Society, Tucson.

Allred, Dorald M., Beck, D Elden, and Jorgensen, Clive D. 1963. "Biotic Communities of the Nevada Test Site," Brigham Young University Science Bulletin, Biological Series II, No. 2.

Anderson, Edgar. 1956. "Man as a Maker of New Plants and New Plant Communities," Man's Role in Changing the Face of the Earth. Chicago: University of Chicago Press, pp. 763–77.

Antevs, Ernst. 1952. "Arroyo-Cutting and Filling," The Journal of Geology, LX, 375–85.

Arizona Citizen, passim. Tucson and Florence.

Arizona Daily Star, passim. Tucson.

Arizona Quarterly Illustrated, July 1880.

Arizona Weekly Enterprise, passim. Florence.

Arizona Weekly Star, passim. Tucson.

Aschmann, Homer. 1959. The Central Desert of Baja California: Demography and Ecology. ("Ibero-Americana," 42.) Berkeley: University of California Press.

Axelrod, Daniel I. 1950. Studies in Late Tertiary Paleobotany. Carnegie Institution of Washington Publication 590. Washington, D. C.: Carnegie Institution.

Baegert, Johann Jakob. 1952. Observations in Lower California. Translated and edited by M. M. Brandenburg and Carl L. Baumann. Berkeley: University of California Press.

Bailey, Reed W. 1935. "Epicycles of Erosion in the Valleys of the Colorado Plateau Province," Journal of Geology, XLIII, 337–55.

Bancroft, Hubert Howe. 1883. The Native Races. Vol. I: Wild Tribes. San Francisco: A. L. Bancroft and Co.

———. 1889. History of the North Mexican States and Texas. 2 vols. San Francisco: The History Co., Publishers.

———. 1962. History of Arizona and New Mexico. 1530–1888. Facsimile of 1889 edition. Albuquerque: Horn and Wallace, Publishers.

Bandelier, A. F. 1892. Final Report of Investigations among the Indians of the Southwestern United States ("Papers of the Archaeological Institute of America," Vol. IV). Cambridge: University Press.

——— (ed.). 1905. The Journey of Alvar Nuñez Cabeza de Vaca. Translated by Fanny Bandelier. New York: A. S. Barnes and Co.

Bannon, John Francis. 1955. The Mission Frontier in Sonora, 1620–1687. New York: U. S. Catholic Historical Society.

Bartlett, John Russell. 1854. Personal Narrative. 2 vols. New York: D. Appleton & Co.

Becker, Carl Lotus. 1932. The Heavenly City of the Eighteenth Century Philosophers. New Haven: Yale University Press.

Bell, William A. 1869. New Tracks in North America. 2 vols. London: Chapman and Hall.

Benson, Lyman. 1950. The Cacti of Arizona. Tucson: The University of Arizona Press.

———. 1962. Plant Taxonomy—Methods and Principles. New York: Ronald Press Co.

———, and Darrow, Robert A. 1954. The Trees and Shrubs of the Southwestern Deserts. Tucson: The University of Arizona Press.

Berber, Laureano Calvo. 1958. Nociones de historia de Sonora. México: Librería de Manuel Porrua.

Berry, James W., and Steelink, Cornelius. 1961. "Chemical Constituents of the Saguaro," Arid Lands Colloquia, 1959–60, 1960–61. The University of Arizona, pp. 39–45.

Bieber, Ralph P. (ed.). 1938. "Cooke's Journal of the March of the Mormon Battalion, 1846–1847." Exploring Southwest Trails, 1846–1854. Vol. VII of Southwest Historical Series. Glendale: Arthur H. Clark Co., pp. 65–240.

Boedecker, L. A. 1930. Letter to Frank Lockwood, December 9, 1930. Tucson, Arizona Pioneers' Historical Society.

Bolton, Herbert Eugene. 1917. "The Mission as a Frontier Institution in the Spanish-American Colonies," American Historical Review, XXIII, 42–61.

——— (trans. and ed.). 1948. Kino's Historical Memoir of Pimeria Alta. Berkeley: University of California Press.

———. 1949. Coronado, Knight of Pueblos and Plains. Albuquerque: University of New Mexico Press.

———. 1960. Rim of Christendom. Reissued. New York: Russell and Russell.

———, and Marshall, Thomas M. 1920. The Colonization of North America, 1492–1783. New York: The Macmillan Co.

Borchert, John R. 1950. "The Climate of the Central North American Grassland," Annals of the Association of American Geographers, XL, 1–39.

Bourke, John Gregory. 1950. On the Border with Crook. Reprinted. Columbus: Long's College Book Co.

Brady, L. F. 1936. "The Arroyo of the Rio de Flag." Museum Notes, Museum of Northern Arizona, IX (Dec.), 33–37.

Brand, Donald D. 1936. "Notes to Accompany a Vegetation Map of Northwest Mexico," University of New Mexico Bulletin No. 280, Biological Series IV, No. 4.

Brandes, Ray. 1960. Frontier Military Posts of Arizona. Globe, Arizona: Dale Stuart King, Publisher.

Branscomb, Bruce L. 1956. "Shrub Invasion of a Southern New Mexico Desert Grassland Range." M.S. thesis, The University of Arizona.

———. 1958. "Shrub Invasion of a Southern New Mexico Desert Grassland Range," Journal of Range Management, XI, 129–32.

Browne, J. Ross. 1951. A Tour through Arizona, 1864. Republished. Tucson: Arizona Silhouettes.

Bryan, Kirk. 1925. "Date of Channel Trenching (Arroyo Cutting) in the Arid Southwest," Science, n.s. LXII, 338–44.

———. 1927. Channel Erosion of the Rio Salado, Socorro County, New Mexico. USGS Bulletin 790, pp. 17–19.

———. 1928a. "Change in Plant Associations by Change in Ground Water Level," Ecology, IX, 474–78.

———. 1928b. "Historic Evidence on Changes in the Channel of Rio Puerco," Journal of Geology, XXXVI, 265–82.

———. 1940. "Erosion in the Valleys of the Southwest," New Mexico Quarterly, X, 227–32.

Bryson, Reid A. 1957. "The Annual March of Precipitation in Arizona, New Mexico, and Northwestern Mexico." Technical Reports on the Meteorology and Climatology of Arid Regions, No. 6, The University of Arizona, Institute of Atmospheric Physics.

———, and Lowry, William P. 1955. "Synoptic Climatology of the Arizona Summer Precipitation Singularity," Bulletin of the American Meteorological Society, XXXVI, 329–39.

Butzer, K. W. 1957. "The Recent Climatic Fluctuation in Lower Latitudes and the General Circulation of the Pleistocene," Geografiska Annaler, XXXIX, 105–13.

Callendar, G. S. 1961. "Temperature Fluctuations and Trends over the Earth," Quarterly Journal of the Royal Meteorological Society, LXXXVII, 1–12.

Calvin, Ross. n.d. "The History of the Upper Gila Region in Arizona and New Mexico." MS in Arizona Pioneers' Historical Society, Tucson.

Cannon, William A. 1916. "Distribution of the Cacti with Especial Reference to the Role Played by the Root Response to Soil Temperature and Soil Moisture," American Naturalist, L, 435–42.

Carson, Christopher. 1926. Kit Carson's Own Story of His Life. Edited by Blanche C. Grant. Taos, New Mexico.

Carter, George F. 1950. "Ecology—Geography—Ethnobotany," Scientific Monthly, LXX, 73–80.

Castetter, E. F. 1956. "The Vegetation of New Mexico," New Mexico Quarterly, XXVI, 257–88.

Chamberlain, William H. 1945. "From Lewisburg to California in 1849." Edited by Lansing B. Bloom. New Mexico Historical Review, XX, 144–80, 239–68.

Chapman, Charles Edward. 1916. The Founding of Spanish California. New York: The Macmillan Co.

Clarke, A. B. 1852. Travels in Mexico and California. Boston: Wright & Hasty's Steam Press.

Cleland, Robert Glass. 1950. This Reckless Breed of Men. New York: Alfred A. Knopf.

Clements, Frederic E. 1920. Plant Indicators: The Relation of Plant Communities to Process and Practice. Carnegie Institution of Washington Publication 290. Washington, D. C.: Carnegie Institution.

Cochise County District Court. 1889. Antonio Grijalba et al. vs. Thos. Dunbar et al., 3 vols. MS in Arizona Pioneers' Historical Society, Tucson.

Cole, L. C. 1957. "Biological Clock in the Unicorn," Science, n.s. CXXV, 874–76.

Colton, Harold S. 1937. "Some Notes on the Original Condition of the Little Colorado River: A Side Light on the Problems of Erosion," Museum Notes, Museum of Northern Arizona, X (Dec.), 17–20.

Conkling, Roscoe P., and Margaret B. 1947. The Butterfield Overland Mail, 1857–1869. 3 vols. Glendale: Arthur H. Clark Co.

Contreras Arias, Alfonso. 1942. Mapas de las Provincias Climatológicas de la República Mexicana. México: Secretaría de Agricultura y Fomento, Dirreción de Geografía, Meteorología e Hidrología.

Cook, Sherburne F. 1940. Population Trends among the California Mission Indians. ("Ibero-Americana," 17.) Berkeley: University of California Press.

———. 1942. "The Population of Mexico in 1793," Human Biology, XIV, 499–515.

———, and Simpson, Lesley Byrd. 1948. The Population of Central Mexico in the Sixteenth Century. ("Ibero-Americana," 31.) Berkeley: University of California Press.

Cottam, W. P., and Stewart, George. 1940. "Plant Succession as a Result of Grazing and of Meadow Desiccation by Erosion since Settlement in 1862," Journal of Forestry, XXXVIII, 613–26.

Coupland, Robert T. 1959. "Effects of Changes in Weather Conditions upon Grasslands in the Northern Great Plains," Grasslands. Washington: American Association for the Advancement of Science.

Couts, Cave Johnson. 1961. Hepah, California. Edited by Henry F. Dobyns. Tucson: Arizona Pioneers' Historical Society.

Cox, C. C. 1925. "From Texas to California in 1849." Edited by Mabelle Martin. Southwestern Historical Quarterly, XXIX, 128–46.

Daily Alta Californian, passim. San Francisco.

Dale, E. E. 1930. The Range Cattle Industry. Norman: University of Oklahoma.

Darrow, R. A. 1944. Arizona Range Resources and Their Utilization: I. Cochise County. The University of Arizona Agricultural Experiment Station Bulletin 103.

Daubenmire, R. F. 1956. "Climate as a Determinant of Vegetation Distribution in Eastern Washington and Northern Idaho," Ecological Monographs, XXVI, 131–54.

———. 1957. "Injury to Plants from Rapidly Dropping Temperature in Washington and Northern Idaho," Journal of Forestry, LV, 581–85.

———. 1959. Plants and Environment. New York: John Wiley & Sons.

Decorme, Gerard. 1941. La obra de los Jesuitas Mexicanos durante la época colonial, 1572–1767. 2 vols. México: José Porrua e Hijos.

Dellenbaugh, F. S. 1912. "Cross Cutting and Retrograding of Stream-Beds," Science, n.s. XXXV, 656–58.

Dickson, Charles Ray. 1958. "Ground Layer Temperature Inversions in an Interior Valley and Canyon." University of Utah, Department of Meteorology.

Di Peso, Charles C. 1953. The Sobaipuri Indians of the Upper San Pedro River Valley, Southeastern Arizona. Dragoon, Arizona: The Amerind Foundation, Inc.

Dittmer, Howard J. 1951. Vegetation of the Southwest —Past and Present," Texas Journal of Science, III, 350–55.

Dorroh, J. H., Jr. 1946. Certain Hydrologic and Climatic Characteristics of the Southwest. University of New Mexico Publications in Engineering, Number One. Albuquerque: University of New Mexico Press.

Duce, James Terry. 1918. "The Effect of Cattle on the Erosion of Canon [sic] Bottoms," Science, n.s. XLVII, 450–52.

Dunne, Peter Masten. 1957. Juan Antonio Balthasar: Padre Visitador to the Sonora Frontier, 1744–1745. Tucson: Arizona Pioneers' Historical Society.

Durivage, John E. 1937. "Journal," Southern Trails to California in 1849. Edited by Ralph P. Bieber. Vol. V of Southwest Historical Series. Glendale: Arthur H. Clark Co., pp. 159–255.

Eaton, W. Clement. 1933. "Frontier Life in Southern Arizona, 1858–61," Southwestern Historical Quarterly, XXXVI, 173–92.

Eccleston, Robert. 1950. Overland to California on the Southwestern Trail: 1849. Edited by George P. Hammond and Edward H. Howes. Berkeley: University of California Press.

Elton, Charles S. 1958. The Ecology of Invasions by Animals and Plants. London: Methuen and Co.

Emory, William Hemsley. 1848. "Notes of a Military Reconnoissance." U. S. Congress, House, Executive Document 41, 30th Cong., 1st Sess. Washington: Wendell and Van Benthuysen, Printers, pp. 15–126.

———. 1857. Report on the United States and Mexican Boundary Survey. 2 vols. U. S. Congress, Senate, Executive Document 108, 34th Cong., 1st Sess. Washington: A. O. P. Nicholson, Printer.

Etz, Della Bohn. 1939. "Reminiscences." MS in Arizona Pioneers' Historical Society, Tucson.

Evans, George W. B. 1945. Mexican Gold Trail: The Journal of a Forty-Niner. Edited by Glenn S. Dumke. San Marino, California: Huntington Library.

Ewing, Russell C. 1934. "The Pima Uprising, 1751–52." Ph.D. dissertation, University of California.

"Extracts from the Journal of Henry W. Bigler." 1932. Utah Historical Quarterly, V, 35–64, 87–112, 134–160.

Forbes, Jack Douglas. 1957. "The Janos, Jocomes, Mansos and Sumas Indians," New Mexico Historical Review, XXXII, 319–34.

Forbes, Robert H. 1958. The Expanding Sahara. The University of Arizona Physical Science Bulletin 3.

Furness, Franklin N. (ed.). 1961. Solar Variations, Climatic Change, and Related Geophysical Problems. (Annals of the New York Academy of Sciences, Vol. 95, Art.1, pp. 1–740.)

Galbraith, F. W. 1959. "Craters of the Pinacates," Guidebook II, Arizona Geological Society. The University of Arizona, pp. 161–64.

Geiger, Rudolf. 1957. The Climate Near the Ground. Translated by Milroy N. Stewart. Cambridge: Harvard University Press.

Gentry, Howard Scott. 1942. Rio Mayo Plants. Carnegie Institution of Washington Publication 527. Washington, D. C.: Carnegie Institution.

Glendening, George E. 1952. "Some Quantitative Data on the Increase of Mesquite and Cactus on a Desert Grassland Range in Southern Arizona," Ecology, XXXIII, 319–28.

———, and Paulsen, Harold A., Jr. 1955. Reproduction

and Establishment of Velvet Mesquite as Related to Invasion of Semidesert Grasslands. USDA, Forest Service Technical Bulletin 1127.

Golder, Frank Alfred, Bailey, Thomas A., and Smith, J. Lyman. 1928. The March of the Mormon Battalion. New York: The Century Co.

Gould, Frank W. 1951. Grasses of Southwestern United States. The University of Arizona, Biological Science Bulletin No. 7.

Green, Christine R. 1959. "Arizona Statewide Rainfall." Technical Reports on the Meteorology and Climatology of Arid Regions, No. 7. The University of Arizona, Institute of Atmospheric Physics.

———. 1962. "Heating and Cooling Degree-Day Characteristics in Arizona." Technical Reports on the Meteorology and Climatology of Arid Regions, No. 10. The University of Arizona, Institute of Atmospheric Physics.

———, and Sellers, William D. (eds.). 1964. Arizona Climate. Tucson: The University of Arizona Press.

Greever, William S. 1957. "Railway Development in the Southwest," New Mexico Historical Review, XXXII, 151–203.

Gregory, Herbert E. 1917. Geology of the Navajo Country. USGS Professional Paper 93.

———, and Moore, Raymond C. 1931. The Kaiparowits Region. USGS Professional Paper 164.

Griffin, John S. 1943. A Doctor Comes to California. San Francisco: California Historical Society.

Griffiths, David. 1904. Range Investigations in Arizona. USDA, Bureau of Plant Industry Bulletin 67.

———. 1910. A Protected Stock Range in Arizona. USDA, Bureau of Plant Industry Bulletin 177.

Hack, John T. 1939. "The Late Quarternary History of Several Valleys of Northern Arizona: A Preliminary Announcement," Museum Notes, Museum of Northern Arizona, XI (May), 67–73.

Haley, J. Evetts (ed.). 1932. "A Log of the Texas-California Cattle Trail, 1854," Southwestern Historical Quarterly, XXXV, 290–316.

Hallenbeck, Cleve. 1940. The Journey and Route of Alvar Núñez, Cabeza de Vaca. Glendale: Arthur H. Clark Co.

Hamilton, Patrick. 1881. The Resources of Arizona. Prescott.

Hammond, George P., and Rey, Agapito (eds.). 1940. Narratives of the Coronado Expedition, 1540–42. Albuquerque: University of New Mexico Press.

Hardy, Robert William Hale. 1829. Travels in the Interior of Mexico in 1825, 1826, 1827, and 1828. London: H. Colburn and R. Bentley.

Haskett, Bert. 1935. "Early History of the Cattle Industry in Arizona," Arizona Historical Review, VI, 3–42.

Hastings, James Rodney. 1959. "Vegetation Change and Arroyo Cutting in Southeastern Arizona," Journal of the Arizona Academy of Science, I, 60–67.

———. 1961a. "People of Reason and Others: the Colonization of Sonora to 1767," Arizona and the West, III, 321–40.

———. 1961b. "Precipitation and Saguaro Growth," Arid Lands Colloquia, 1959–60, 1960–61. The University of Arizona, pp. 30–38.

———. 1964a. "Climatological Data for Baja California." Technical Reports on the Meteorology and Climatology of Arid Regions, No. 14. The University of Arizona, Institute of Atmospheric Physics.

———. 1964b. "Climatological Data for Sonora and Northern Sinaloa." Technical Reports on the Meteorology and Climatology of Arid Regions, No. 15. The University of Arizona, Institute of Atmospheric Physics.

———. 1965. "On Some Uses of Nonnormal Coefficients of Variation," Journal of Applied Meterology (in press).

———, and Alcorn, Stanley M. 1961. "Physical Determinations of Growth and Age in the Giant Cactus," Journal of the Arizona Academy of Science, II, 32–39.

———, and Turner, Raymond M. 1965. "Seasonal Precipitation Regimes in Baja California" (forthcoming).

Haury, Emil W. 1958. "Post-Pleistocene Human Occupation of the Southwest," Climate and Man in the Southwest. Tucson: The University of Arizona Press.

Hinton, Richard J. 1890. Irrigation in the United States. Vol. IV of Report of the Special Committee of the United States Senate on the Irrigation and Reclamation of Arid Lands. Senate Report 928, 51st Cong., 1st Sess. Washington: Government Printing Office.

———. 1954. The Handbook to Arizona. Republished. Tucson: Arizona Silhouettes.

History of Arizona Territory. 1884. San Francisco: Wallace W. Elliott and Co.

Hitchcock, A. S. 1950. Manual of the Grasses of the United States. 2d ed. USDA Miscellaneous Publication 200.

Hornaday, W. T. 1908. Camp-Fires on Desert and Lava. New York: Charles Scribner's Sons.

Hubbs, Carl L. 1957. "Recent Climatic History in California and Adjacent Areas," Proceedings of the Conference of Recent Research in Climatology. Scripps Institute of Oceanography, pp. 10–22.

Humphrey, Robert R. 1949. "Fire as a Means of Controlling Velvet Mesquite, Burroweed, and Cholla on Southern Arizona Ranges," Journal of Range Management, II, 175–82.

———. 1953. "The Desert Grassland, Past and Present," Journal of Range Management, VI, 159–64.

———. 1958. "The Desert Grassland, a History of Vegetational Change and an Analysis of Causes," The Botanical Review, XXIV, 193–252.

———. 1960. Forage Production on Arizona Ranges V: Pima, Pinal and Santa Cruz Counties. The University of Arizona Agricultural Experiment Station Bulletin 302.

———. 1962. Range Ecology. New York: Ronald Press Company.

Huntington, Ellsworth. 1914. The Climatic Factor. Carnegie Institution of Washington Publication 192. Washington, D. C.: Carnegie Institution.

Huxley, Julian. 1961. Of Wild Life and Natural Habitats in Central and East Africa. Paris: UNESCO.

Huzayyin, Soliman. 1956. "Changes in Climate, Vegetation, and Human Adjustment in the Saharo-Arabian Belt with Special Reference to Africa," Man's Role in Changing the Face of the Earth. Chicago: University of Chicago Press, pp. 304–23.

Itinerary of the El Paso and Fort Yuma Wagon Road Expedition under the Superintendence of James B. Leach. 1858. Film Microcopies of Records in the National Archives, No. 95, Roll 3. "Records of the Office of the Secretary of the Interior Relating to Wagon Roads, 1857–1881."

Ives, Ronald L. 1949. "Climate of the Sonoran Desert Region," Annals of the Association of American Geographers, XXXIX, 143–87.

Jahns, Richard H. 1959. "Collapse Depressions of the Pinacate Volcanic Field, Sonora, Mexico," Guidebook II, Arizona Geological Society. The University of Arizona, pp. 165–83.

Johnson, Donald E. 1961. "Edaphic Factors Affecting the Distribution of Creosotebush, *Larrea tridentata* (DC) Cov., in Desert Grassland Sites of Southeastern Arizona." M. S. thesis, The University of Arizona.

Johnston, A. R. 1848. "Journal." U. S. Congress, House, Executive Document 41, 30th Cong., 1st Sess. Washington: Wendell and Van Benthuysen, Printers, pp. 565–614.

"The Journal of Nathaniel V. Jones, With the Mormon Battalion." 1931. Utah Historical Quarterly, IV, 6–23.

"The Journal of Robert S. Bliss, With the Mormon Battalion." 1931. Utah Historical Quarterly, IV, 67–96, 110–28.

Judson, Sheldon. 1952. "Arroyos," Scientific American, CLXXXVII, 71–76.

Kearney, Thomas H., Peebles, Robert H., *et al.* 1960. Arizona Flora. 2d ed. Berkeley: University of California Press.

Keleher, William A. 1952. Turmoil in New Mexico. Santa Fe: The Rydal Press.

Kincer, J. B. 1933. "Is Our Climate Changing?" Monthly Weather Review, LXI, 251–59.

——. 1946. "Our Changing Climate," Transactions of the American Geophysical Union, XXVII, 342–47.

Lafora, Nicolás de. 1939. Relación del viaje que hizo a los presidios internos situados en la frontera de la América septentrional. México: Editorial Pedro Robredo.

Lake, J. V. 1956. "The Temperature Profile above Bare Soil on Clear Nights," Quarterly Journal of the Royal Meteorological Society, LXXXII, 187–97.

Land, Edward. 1934. "Reminiscences," MS in Arizona Pioneers' Historical Society, Tucson.

Landsberg, H. E., and Mitchell, J. M., Jr. 1961. "Temperature Fluctuations and Trends over the Earth," Quarterly Journal of the Royal Meteorological Society, LXXXVII, 435–36.

Lavender, David. 1954. Bent's Fort. Garden City: Doubleday & Co.

Leopold, Luna B. 1951a. "Rainfall Frequency: an Aspect of Climatic Variation," Transactions, American Geophysical Union, XXXII, 347–57.

——. 1951b. "Vegetation of Southwestern Watersheds in the Nineteenth Century," The Geographical Review, XLI, 295–316.

——, Leopold, E. B., and Wendorf, F. 1963. "Some Climatic Indicators in the Period A.D. 1200–1400 in New Mexico," Changes of Climate. UNESCO, pp. 265–70.

——, and Miller, John P. 1954. A Postglacial Chronology for Some Alluvial Valleys in Wyoming. USGS Water-Supply Paper 1261.

——, and Snyder, C. T. 1951. Alluvial Fills near Gallup, New Mexico. USGS Water-Supply Paper 1110-A.

Life, August 18, 1952.

Lockwood, Frank C. 1929. "American Hunters and Trappers in Arizona," Arizona Historical Review, II, 70–85.

Love, Clara M. 1916. "History of the Cattle Industry in the Southwest," Southwestern Historical Quarterly, XIX, 370–99; XX, 1–18.

Lumholtz, Carl. 1912. New Trails in Mexico. New York: Charles Scribner's Sons.

MacDougal, D. T. 1908. "Across Papagueria." American Geographical Society Bulletin, XL, 705–25.

Malin, James C. 1956. The Grassland of North America. Lawrence, Kansas: James C. Malin.

Mange, Juan Matheo. 1926. Luz de tierra incógnita en la América septentrional y diario de las exploraciones en Sonora. Edited by Francisco del Castillo. México: Talleres Gráficos de la Nación.

"Man's Tenure of the Earth." 1956. Man's Role in Changing the Face of the Earth. Chicago: University of Chicago Press, pp. 401–9.

Mann, Dean E. 1963a. The Politics of Water in Arizona. Tucson: The University of Arizona Press.

——. 1963b. "Political and Social Institutions in Arid Regions," Aridity and Man. Washington: American Association for the Advancement of Science, 397–428.

Marks, John Brady. 1950. "Vegetation and Soil Relations in the Lower Colorado Desert," Ecology, XXXI, 176–93.

Marshall, Joe T., Jr. 1957. Birds of Pine-Oak Woodland in Southern Arizona and Adjacent Mexico. Berkeley: Cooper Ornithological Society.

Martin, Paul S. 1963. The Last 10,000 Years. Tucson: The University of Arizona Press.

——, Schoenwetter, James, and Arms, Bernard C. 1961. The Last 10,000 Years. The University of Arizona Geochronology Laboratories.

Matson, Daniel S., and Schroeder, Albert H. (eds.). 1957. "Cordero's Description of the Apache—1796," New Mexico Historical Review, XXXII, 335–56.

Mattison, Ray H. 1946. "Early Spanish and Mexican Settlements in Arizona," New Mexico Historical Review, XXI, 273–327.

McDonald, James E. 1956. "Variability of Precipitation in an Arid Region: a Survey of Characteristics for Arizona." Technical Reports on the Meteorology and Climatology of Arid Regions, No. 1. The University of Arizona, Institute of Atmospheric Physics.
————. 1959. "Climatology of Arid Lands," Arid Lands Colloquia, 1958–59. The University of Arizona, pp. 3–13.
————. 1962. "The Evaporation-Precipitation Fallacy," Weather, XVII, 1–9.
McGee, W. J. 1898. "The Seri Indians," Annual Report of the Bureau of Ethnology No. 17, 1895–96. Washington: Government Printing Office, pp. 9–344.
McGregor, S. E., Alcorn, Stanley M., and Olin, George. 1962. "Pollination and Pollinating Agents of the Saguaro." Ecology, XLIII, 259–67.
Mehrhoff, L. A., Jr. 1955. "Vegetation Changes on a Southern Arizona Grassland Range—an Analysis of Causes," M.S. thesis, The University of Arizona.
Meteorological Abstracts and Bibliography. 1950. I, 473–75.
Mitchell, J. Murray, Jr. 1953. "On the Causes of Instrumentally Observed Secular Temperature Trends," Journal of Meteorology, X, 244–61.
————. 1961. "Bibliographic List of Recent Studies of Climatic Change by United States Citizens." USDC, Weather Bureau.
————. 1963. "On the World-Wide Pattern of Secular Temperature Change," Changes of Climate. UNESCO, pp. 161–81.
Morello, Jorge H., and Saravia Toledo, Carlos. 1959. "El Bosque Chaqueño. I. Paisaje Primitivo, Paisaje Natural y Paisaje Cultural en el Oriente de Salta," Revista Agronómica del Noroeste Argentino, III, 5–81.
Morgan, Willard D., and Lester, Henry M. 1954. Graphic Graflex Photography. 10th ed. New York: Morgan and Lester.
Morris, Ralph C. 1926. "The Notion of a Great American Desert East of the Rockies," Mississippi Valley Historical Review, XIII, 190–200.
Morrisey, Richard J. 1950. "The Early Range Cattle Industry in Arizona," Agricultural History, XXIV, 151–56.
————. 1951. "The Northward Expansion of Cattle Ranching in New Spain, 1550–1600," Agricultural History, XXV, 115–21.
Muller, Cornelius H. 1947. "Vegetation and Climate of Coahuila, Mexico," Madroño IX, 33–57.
Murphy, Frank. 1928. "Reminiscences," MS in Arizona Pioneers' Historical Society, Tucson.

Nelson, E. W. 1934. The Influence of Precipitation and Grazing upon Black Grama Grass Range. USDA Technical Bulletin 409.
[Nentuig, Juan.] 1951. Rudo Ensayo. Translated by Eusebio Guiteras. Republished. Tucson: Arizona Silhouettes.
Nichol, A. A. 1952. The Natural Vegetation of Arizona. The University of Arizona Agricultural Experiment Station Technical Bulletin 127.

Niering, W. A., Whittaker, R. H., and Lowe, C. H. 1963. "The Saguaro: a Population in Relation to Environment," Science CXLII, 15–23.
Norris, J. J. 1950. Effect of Rodents, Rabbits, and Cattle on Two Vegetation Types in Semidesert Range Land. New Mexico Agricultural Experiment Station Bulletin 353.

Ohnesorgen, William. 1929. "Reminiscences," MS in Arizona Pioneers' Historical Society, Tucson.
Olmstead, F. H. 1919. "A Report on Flood Control of the Gila River in Graham County, Arizona." U. S. Congress, Senate, Senate Document 436, 65th Cong., 3d Sess. Washington: Government Printing Office.
Osgood, E. S. 1929. The Day of the Cattleman. Minneapolis: University of Minnesota.

Page, John L. 1930. "Climate of Mexico," Monthly Weather Review, Supplement No. 33. Washington: Government Printing Office.
Page, L. F. 1937. "Temperature and Rainfall Changes in the United States during the Past Forty Years," Monthly Weather Review, LXV, 46–55.
Parke, John G. 1857. Report of Explorations for Railroad Routes. Vol. VII of Reports of Explorations and Surveys. U. S. Congress, House, House Executive Document 91, 33d Cong., 2d Sess. Washington: A.O.P. Nicholson, Printer.
Parker, Kenneth W., and Martin, S. Clark. 1952. The Mesquite Problem on Southern Arizona Ranges. USDA Circular 908.
Pattie, James O. 1905. The Personal Narrative of James O. Pattie, of Kentucky. Vol. XVIII of Early Western Travels. Edited by Reuben Gold Thwaites. Cleveland: Arthur H. Clark Co.
Paulsen, Harold A., Jr. 1950. "Mortality of Velvet Mesquite Seedlings," Journal of Range Management, III, 281–86.
Pelzer, Louis. 1936. The Cattlemen's Frontier. Glendale: Arthur H. Clark Co.
Pérez de Ribas, Andrés. 1645. Historia de los triunfos de nuestra Santa Fé entre gentes las mas bárbaras, y fieras del Nuevo Orbe. Madrid: A. de Paredes.
Peterson, H. V. 1950. "The Problem of Gullying in Western Valleys," Applied Sedimentation. Edited by Parker D. Trask. New York: John Wiley and Sons, pp. 407–34.
Pfefferkorn, Ignaz. 1949. Sonora, a Description of the Province. Translated and edited by Theodore E. Treutlein. Albuquerque: The University of New Mexico Press.
Phillips, Frank J. 1912. Emory Oak in Southern Arizona. USDA, Forest Service Circular 201.
Phillips, Walter S. 1963. Vegetational Changes in Northern Great Plains. The University of Arizona Agricultural Experiment Station Report 214.
Pool, Frank. 1935. "Reminiscences." MS in Arizona Pioneers' Historical Society, Tucson.
Powell, H. M. T. 1931. The Santa Fe Trail to California, 1849–1852. San Francisco: Grabhorn Press.
Powell, John Wesley. 1962. Report on the Lands of the Arid Region of the United States. Edited by Wal-

lace Stegner. Cambridge: The Belknap Press of Harvard University Press.

Pradeau, Alberto Francisco. 1953. "Nentuig's 'Description of Sonora.'" Mid-America, XXXV, pp. 81–90.

Priestley, Herbert I. 1916. José de Gálvez, Visitor-General of New Spain, 1765–1771. Berkeley: University of California Press.

Pumpelly, Raphael. 1870. Across America and Asia. New York: Leypoldt and Holt.

Reagan, Albert B. 1924. "Recent Changes in the Plateau Region," Science, n.s. LX, 283–85.

Report of the Attorney General. 1904. Washington: Government Printing Office.

Report of the Governor of Arizona to the Secretary of the Interior. 1885. Washington: Government Printing Office.

Report of the Governor of Arizona to the Secretary of the Interior. 1896. Washington: Government Printing Office.

Report on Barracks and Hospitals with Descriptions of Military Posts. 1870. Circular No. 4, War Department, Surgeon General's Office. Washington: Government Printing Office.

Reyes, Antonio María de los. 1938. "Memorial sobre las Misiones de Sonora, 1772," Boletín del Archivo General de la Nación, IX, 276–320.

———. 1958. Relación hecha el año de 1784 de las Misiones establecidas en Sinaloa y Sonora. Mexico.

Reynolds, Hudson G. 1950. "Relation of Merriam Kangaroo Rats to Range Vegetation in Southern Arizona," Ecology, XXXI, 456–63.

———, and Bohning, J. W. 1956. "Effects of Burning on a Desert Grass-Shrub Range in Southern Arizona," Ecology XXXVII, 769–77.

———, and Glendening, George E. 1949. "Merriam Kangaroo Rat a Factor in Mesquite Propagation on Southern Arizona Range Lands," Journal of Range Management, II, 193–97.

Rich, John Lyon. 1911. "Recent Stream Trenching in the Semi-Arid Portion of Southwestern New Mexico, a Result of Removal of Vegetation Cover," American Journal of Science, XXXII, 237–45.

Rothrock, J. T. 1875. "Preliminary and General Botanical Report, with Remarks upon the General Topography of the Region Traversed." Annual Report of the Chief of Engineers, 1875. Washington: Government Printing Office.

Russell, Frank. 1908. "The Pima Indians," Annual Report of the Bureau of American Ethnology No. 26, 1904–05. Washington: Government Printing Office, pp. 4–390.

Sauer, Carl O. 1932. The Road to Cibola. ("Ibero-Americana," 3.) Berkeley: University of California Press.

———. 1934. The Distribution of Aboriginal Tribes and Languages in Northwestern Mexico. ("Ibero-Americana," 5.) Berkeley: University of California Press.

———. 1935. Aboriginal Population of Northwestern Mexico. ("Ibero-Americana," 10.) Berkeley: University of California Press.

———. 1944. "A Geographic Sketch of Early Man," Geographical Review, XXXIV, 529–73.

———. 1956. "The Agency of Man on the Earth." Man's Role in Changing the Face of the Earth. Chicago: University of Chicago Press, pp. 49–69.

———, and Brand, Donald. 1931. "Prehistoric Settlements of Sonora," University of California Publications in Geography, V, 67–148.

Schrader, Frank C. 1915. Mineral Deposits of the Santa Rita and Patagonia Mountains, Arizona. USGS Bulletin 582.

Schroeder, Albert R. 1956. "Fray Marcos de Niza, Coronado and the Yavapai," New Mexico Historical Review, XXXI, 24–37.

Schulman, Edmund. 1956. Dendroclimatic Changes in Semiarid America. Tucson: The University of Arizona Press.

Schumm, S. A., and Hadley, R. F. 1957. "Arroyos and the Semiarid Cycle of Erosion," American Journal of Science, CCLV, 161–74.

Schwennesen, A. T. 1917. Ground Water in San Simon Valley, Arizona and New Mexico. USGS Water-Supply Paper 425A.

Secretaría de Recursos Hidráulicos. 1954. Boletín Hidrológico Número 11. México.

———. 1959. Boletín Hidrológico Número 13. México.

Sears, Paul B. 1947. Deserts on the March. Norman: University of Oklahoma Press.

Sellers, William D. 1960. "Precipitation Trends in Arizona and Western New Mexico," Proceedings of the 28th Annual Snow Conference, Santa Fe, April, pp. 81–94.

Shantz, H. L. 1905. "A Study of the Vegetation of the Mesa Region East of Pike's Peak: The Bouteloua Formation," Botanical Gazette, XLII, 16–47, 179–207.

———. 1924. Natural Vegetation: Grassland and Desert Shrub. USDA, Atlas of American Agriculture, pp. 15–29.

———, and Piemeisel, R. L. 1924. "Indicator Significance of the Natural Vegetation of the Southwestern Desert Region," Journal of Agricultural Research, XXVIII, 721–801.

———, and Turner, B. L. 1958. Vegetational Changes in Africa. The University of Arizona, College of Agriculture Report 169.

Shapley, Harlow. 1953. Climatic Change: Evidence, Causes, and Effects. Cambridge: Harvard University Press.

Shreve, Forrest. 1910. "The Rate of Establishment of the Giant Cactus," Plant World, XIII, 235–40.

———. 1911a. "Establishment Behavior of the Palo Verde," Plant World, XIV, 289–96.

———. 1911b. "The Influence of Low Temperatures on the Distribution of the Giant Cactus," Plant World, XIV, 136–46.

———. 1912. "Cold Air Drainage," Plant World, XV, 110–15.

Shreve, Forrest. 1915. The Vegetation of a Desert Mountain Range as Conditioned by Climatic Factors. Carnegie Institution of Washington Publication 217. Washington, D. C.: Carnegie Institution.

———. 1917a. "The Establishment of Desert Perennials," Journal of Ecology, V, 210–16.

———. 1917b. "A Map of the Vegetation of the United States," Geographical Review, III, 119–25.

———. 1922. "Conditions Indirectly Affecting Vertical Distribution on Desert Mountains," Ecology, III, 269–74.

———. 1925. "Ecological Aspects of the Deserts of California," Ecology, VI, 93–103.

———. 1934a. "Rainfall, Runoff, and Soil Moisture under Desert Conditions," Annals of the Association of American Geographers, XXIV, 131–56.

———. 1934b. "Vegetation of the Northwestern Coast of Mexico," Bulletin of the Torrey Botanical Club, LXI, 373–80.

———. 1939. "Observations on the Vegetation of Chihuahua,"Madroño,V,1–13.

———. 1942a. "The Desert Vegetation of North America," The Botanical Review, VIII, 195–246.

———. 1942b. "Grassland and Related Vegetation in Northern Mexico," Madroño, VI, 190–98.

———. 1942c. "The Vegetation of Arizona," in Flowering Plants and Ferns of Arizona by T. H. Kearney and R. H. Peebles. USDA Miscellaneous Publications 423, pp. 10–23.

———. 1944. "Rainfall of Northern Mexico," Ecology, XXV, 105–11.

———. 1964. Vegetation of the Sonoran Desert, in Vegetation and Flora of the Sonoran Desert by Forrest Shreve and Ira L. Wiggins. 2 vols. Stanford, California: Stanford University Press.

———, and Hinckley, Arthur L. 1937. "Thirty Years of Change in Desert Vegetation," Ecology, XVIII, 463–78.

Simpson, Lesley Byrd. 1952. Exploitation of Land in Central Mexico in the Sixteenth Century. ("Ibero-Americana," 36.) Berkeley: University of California.

Sinclair, John G. 1922. "Temperatures of the Soil and Air in a Desert," Monthly Weather Review, L, 142–44.

Southwestern Stockman, passim. Willcox, Arizona.

Southwest Watershed Hydrology Studies Group. 1958. Annual Progress Report. Tucson and Tombstone, Arizona: Project Offices.

Spicer, Edward H. 1962. Cycles of Conquest. Tucson: The University of Arizona Press.

Spring, John. 1902. "With the Regulars in Arizona in the Sixties," Washington National Tribune, November 20.

———. 1903. "Troublous Days in Arizona," Washington National Tribune, July-October.

Stamp, L. Dudley. 1961. "Some Conclusions," A History of Land Use in Arid Regions. Paris: UNESCO, pp. 379–88.

Standley, Paul C. 1920–26. Trees and Shrubs of Mexico. 5 parts. Washington: Smithsonian Institution.

Stevens, Robert C. 1964. "The Apache Menace in Sonora," Arizona and the West, VI, 211–22.

Stewart, Omer C. 1951. "Burning and Natural Vegetation in the United States," Geographical Review, XLI, 317–20.

———. 1956. "Fire as the First Great Force Employed by Man," Man's Role in Changing the Face of the Earth. Chicago: University of Chicago Press, pp. 115–33.

Sutton, O. G. 1953. Micrometeorology. New York: McGraw-Hill Book Co.

Swain, Charles H. 1893. "Report on the Mines Known as 'The Old Mowry Mines.' " MS in Arizona Pioneers' Historical Society, Tucson.

Swift, T. T. 1926. "Date of Channel Trenching in the Southwest," Science, n.s. LXIII, 70–71.

Sykes, Godfrey. 1931. "Rainfall Investigations in Arizona and Sonora by Means of Long-Period Rain Gauges," Geographical Review, XXI, 229–33.

Talbot, William J. 1961. "Land Utilization in the Arid Regions of Southern Africa, Part I: South Africa," A History of Land Use in Arid Regions. Paris: UNESCO, pp. 299–331.

Tamarón y Romeral, Pedro. 1937. Demostración del vastísimo obispado de la Nueva Vizcaya, 1765. México: Antigua Librería Robredo de José Porrua e Hijos.

Taylor, Walter P. 1936. "Some Effects of Animals on Plants," Scientific Monthly, XLIII, 262–71.

———, Vorhies, Charles T., and Lister, P. B. 1935. "The Relation of Jack Rabbits to Grazing in Southern Arizona," Journal of Forestry, XXXIII, 490–98.

Tevis, James H. 1954. Arizona in the 50's. Albuquerque: University of New Mexico Press.

Thomas, Alfred Barnaby. 1932. Forgotten Frontiers: A Study of the Spanish Indian Policy of Don Juan Bautista de Anza, Governor of New Mexico, 1777–1787. Norman: University of Oklahoma Press.

———. 1933. "A Description of Sonora in 1772," Arizona Historical Review, V, 302–7.

———. 1941. Teodoro de Croix and the Northern Frontier of New Spain, 1776–1783. Norman: University of Oklahoma Press.

Thornber, J. J. 1910. The Grazing Ranges of Arizona. The University of Arizona Agricultural Experiment Station Bulletin 65.

Thornthwaite, C. Warren. 1948. "An Approach toward a Rational Classification of Climate," Geographical Review, XXXVIII, 55–94.

———, Sharpe, C. F. Stewart, and Dosch, Earl F. 1942. Climate and Accelerated Erosion in the Arid and Semi-Arid Southwest, with Special Reference to the Polacca Wash Drainage Basin, Arizona. USDA Technical Bulletin 808.

Thwaites, Reuben Gold (ed.). 1905. Expedition from Pittsburgh to the Rocky Mountains. Vol. XVII of Early Western Travels. Cleveland: The Arthur H. Clark Co.

The Tombstone, 1885. Tombstone, Arizona.

Tombstone Daily Epitaph, *passim*. Tombstone, Arizona.

Trewartha, Glenn T. 1954. An Introduction to Climate. New York: McGraw-Hill Book Co.

Tschirley, Fred H., and Wagle, R. F. 1964. "Growth Rate and Population Dynamics of Jumping Cholla (*Opuntia fulgida* Engelm.)," Journal of the Arizona Academy of Science, III, 67–71.

Turnage, William V., and Hinckley, A. L. 1938. "Freezing Weather in Relation to Plant Distribution in the Sonoran Desert," Ecological Monographs, VIII, 529–50.

———, and Mallery, T. D. 1941. An Analysis of Rainfall in the Sonoran Desert and Adjacent Territory. Carnegie Institution of Washington Publication 529. Washington, D. C.: Carnegie Institution.

Turner, Raymond M. 1963. "Growth in Four Species of Sonoran Desert Trees," Ecology, XLIV, 760–65.

———, Alcorn, Stanley M., Olin, George, and Booth, John A. 1965. "The Effect of Shade on Saguaro Seedling Establishment," (forthcoming).

Undreiner, George J. 1947. "Fray Marcos de Niza and His Journey to Cibola," The Americas, III, 415–86.

U. S. Bureau of the Census. 1872. Ninth Census of the United States: 1870.

———. 1883. Tenth Census of the United States: 1880.

———. 1895. Eleventh Census of the United States: 1890.

U. S. Congress, Senate. 1852. Senate Executive Document 121, 32d Cong., 1st Sess. Washington: A. Boyd Hamilton.

U. S. Congress, House. 1859. House Executive Document 108, 35th Cong., 2d Sess. Washington: James B. Steedman.

U. S. Congress, Senate. 1880. Senate Executive Document 207, 46th Cong., 2d Sess. Washington: Government Printing Office.

U. S. Congress, Senate. 1898. Report of the Boundary Commission upon the Survey and Re-Marking of the Boundary between the United States and Mexico West of the Rio Grande, 1891 to 1896. Senate Document 247, 55th Cong., 2d Sess. Washington: Government Printing Office.

U. S. Department of Commerce, Weather Bureau. 1965. Substation History, Arizona, Washington, D. C.: United States Government Printing Office.

Velasco, José Francisco. 1850. Noticias estadísticas del estado de Sonora. México: Imprenta de Ignacio Cumplido.

Veryard, R. G. 1963. "A Review of Studies on Climatic Fluctuations during the Period of the Meteorological Record," Changes of Climate. UNESCO, pp. 3–16.

Vivó, Jorge A., and Gómez, José C. 1946. Climatología de México. Mexico: Instituto Panamericano de Geografía e Historia.

von Eschen, G. F. 1958. "Climatic Trends in New Mexico," Weatherwise, XI, 191–95.

Vorhies, Charles T., and Taylor, Walter P. 1933. The Life Histories and Ecology of Jack Rabbits, *Lepus alleni* and *Lepus californicus* ssp., in Relation to Grazing in Arizona. The University of Arizona Agricultural Experiment Station Technical Bulletin Number 49, pp. 467–587.

Wagoner, J. J. 1951. "Development of the Cattle Industry in Southern Arizona, 1870's and 80's," New Mexico Historical Review, XXVI, 204–224.

———. 1952. History of the Cattle Industry in Southern Arizona, 1540–1940. The University of Arizona Social Science Bulletin No. 20.

Wallén, C. C. 1955. "Some Characteristics of Precipitation in Mexico," Geografiska Annaler, XXXVII, 51–85.

Wallmo, O. C. 1955. "Vegetation of the Huachuca Mountains, Arizona," American Midland Naturalist, LIV, 466–80.

Weaver, J. E. 1954. North American Prairie. Lincoln: Johnsen Publishing Co.

Webb, Walter Prescott. 1931. The Great Plains. New York: Ginn and Co.

———. 1957. "The American West—Perpetual Mirage," Harper's Magazine, CCXIV, 25–31.

Weekly Arizonan, 1859. Tubac, Arizona.

Went, F. W. 1949. "Ecology of Desert Plants, II. The Effect of Rain and Temperature on Germination and Growth," Ecology, XXX, 1–13.

———. 1957. The Experimental Control of Plant Growth. Waltham, Mass.: Chronica Botanica Co.

———, and Westergaard, M. 1949. "Ecology of Desert Plants, III. Development of Plants in the Death Valley National Monument, California," Ecology, XXX, 26–38.

West, Robert C. 1949. The Mining Community in Northern New Spain. ("Ibero-Americana," 30.) Berkeley: University of California Press.

White, Stephen S. 1948. "The Vegetation and Flora of the Region of the Río de Bavispe in Northeastern Sonora, Mexico," Lloydia, XI, 229–302.

Whitfield, C. J., and Anderson, H. L. 1938. "Secondary Succession in the Desert Plains Grassland," Ecology, XIX, 171–80.

———, and Beutner, E. L. 1938. "Natural Vegetation in the Desert Plains Grassland," Ecology, XIX, 26–37.

Whyte, R. O. 1963. "The Significance of Climatic Change for Natural Vegetation and Agriculture," Changes of Climate. UNESCO, pp. 381–86.

Wiggins, Ira L. 1964. Flora of the Sonoran Desert, in Vegetation and Flora of the Sonoran Desert by Forrest Shreve and Ira L. Wiggins. 2 vols. Stanford, California: Stanford University Press.

Willett, H. C. 1950. "Temperature Trends of the Past Century," Centenary Proceedings of the Royal Meteorological Society. London: Royal Meteorological Society, pp. 195–206.

Winn, Fred. 1926. "The West Fork of the Gila River," Science, n.s. LXIV, 16–17.

Worchester, Donald E. 1941. "The Beginnings of the Apache Menace of the Southwest," New Mexico Historical Review, XVI, 1–14.

Wyllys, Rufus K. 1931. "Padre Luis Velarde's *Relación* of Pimería Alta, 1716," New Mexico Historical Review, VI, 111–57.

Yang, Tien Wei, and Lowe, Charles H., Jr. 1956. "Correlation of Major Vegetation Climaxes with Soil Characteristics in the Sonoran Desert," Science, CXXIII, 542.

Young, Floyd D. 1921. "Nocturnal Temperature Inversions in Oregon and California," Monthly Weather Review, XLIX, 138–48.

Zohary, Michael. 1962. Plant Life of Palestine. New York: Ronald Press.

Zuñiga, Ignacio. 1835. Rápida ojeada al estado de Sonora. México: Juan Ojeda.

INDEX

The letters "a" and "b" used after page references refer to the left and right columns respectively.

Acacia vernicosa, increase in: 6a, 182, 273
   cattle as agent in spread of: 275
Apache Indians: 24a, 27b
   raids: 29
Aridity *see* Climate; Desert; Precipitation
Arizona Upland province: 185, 187–88
Arizona white oak: 50a
Arroyo cutting: 15a, 41–42, 45
   in Sonora: 31b–32a
   and climate: 43b, 44
   and overgrazing, relation of: 31b, 43–44
   geographical extent of: 45
   and vegetation change: 288a
   plates: 62, 65, 81, 99, 141, 145, 156–57, 159, 163, 175
   see also names of rivers

Babocomari Creek, channel cutting after 1890: 3a
Bajadas, vegetation of: 185, 187a
Barrel cactus, possible destruction by food gatherers: 25a
Beaver: 35b
Blue paloverde: 189a
   on bajadas: 187
   changes in: 270b
   decline in Pinacate region: 271a
   increase in: 273a
   population census, plates: 218, 237
Boojum tree: 189b
Burroweed, increase of 1905–1910: 3b
Bursage: 187a, 188b

Cactus fruit, Indian consumption of: 25b
Canyon forest: 49b
Cardón: 189b, 270b
Catclaw, increase in: 273b
Cattle:
   northward expansion under Spain and Mexico: 33
   as agent in spread of mesquite and woody plants: 6a, 34b, 275–76
   increase of in American period: 40b–41a
   overstocking of: 41

Cattle (*Continued*)
   effect of drought of 1890's on: 41b
   relation to arroyo cutting: 43, 45
   not primary agent of vegetation change: 285b
Cattle, Spanish:
   increase and spread of: 30b–31
   effect of Indian raiding on: 31
Cattle, Spanish and Mexican:
   and arroyo-cutting hypothesis: 31b–32a
   populations: 31a, 33b–34a
   ecological impact of: 31, 34b, 284–85
Cattle, wild: 34
Cattle and climate as causes of change: 289
Cattlemen's associations: 41
Central Gulf Coast province *see* Gulf Coast
Century plants, possible destruction by food gatherers: 25b
Changing mile: 21b
Channel changes *see* Arroyo cutting; Meandering
Channeling, discontinuous: 44a
Chaparral, plates: 112, 113
Chihuahuan Desert: 8b
   plant invaders from: 182b
Cholla, response to temperature: 16a
Cienegas: 26, 37
Climate:
   and vegetation change: 6b
   and vegetation, interaction of: 10
   as determined by vegetation: 10b
   relation to vegetation: 10b, 272b
   and arroyo cutting: 44, 45b
   shift, to drier and hotter conditions: 271a
   records, use of: 279
   and upward displacement of vegetation: 287–88
   see also Meteorology; Precipitation; Temperature
Climate and cattle as causes of change: 289
Climatic change associated with vegetation change: 288
Cold air drainage and plant life: 17b–18a
Colorado River Valley:
   precipitation and temperature: 188a
   vegetation: 188b

Cottonwood, increase in: 274
Cow-flap hypothesis: 6a
Creosote bush:
   as common denominator in deserts: 8b
   on desert bajadas: 187b
   in Colorado River Valley: 188b
   decline in Pinacate region: 271b
Cultivation, ecological impact of: 26
Cyclonic systems: 12

Desert:
   microclimates: 7
   definition of: 7–8, 10a
   world types: 8a
   North American: 8–9
   and climate: 10b
Desert grassland *see* Grassland life zone
Desert life zone: 20a
   soils: 185
   regions of: 185
   variety of vegetation: 187–88
   variability of changes in: 270
   apparent trends in: 270b
   decline of various plant species: 271–73
   see also Arizona Upland province; Colorado River Valley; Gulf Coast
Desert mountains: 10a
Desert region: vii
   utilization of: 4b
   oasis civilization in: 5
   see also Desert; Grasslands; Oak Woodland
"Desertification": 279a
Drought of 1890's: 41b
Dryness and desert: 7a

Ecology, regional:
   impact of Indian food gathering: 24–25
   impact of pre-Hispanic cultivation: 26
   impact of Indians: 27b–28a
   impact of Spanish rule: 30b
   impact of Spanish cattle: 31b
   impact of Spanish mining: 32
   impact of Spanish and Mexican cattle: 34b
   effect of overgrazing: 41b
   impact of Anglo-Americans: 45–46
   impact of rodents: 275–76
   see also Vegetation change